D0152756

Electronic synthesis of speech

This book is dedicated to
my mother and father

Electronic synthesis of speech

R. LINGGARD

Rank-Xerox Professor of Information Technology, Department of Computer Studies, Hull University

'The invention of a talking machine, and its operation in accordance with a well-considered plan, would be one of the boldest schemes to occur to the human intellect.'
Wolfgang von Kempelen 1791

Cambridge University Press

Cambridge
London New York New Rochelle
Melbourne Sydney

Published by the Press Syndicate of the University of Cambridge
The Pitt Building, Trumpington Street, Cambridge CB2 1RP
32 East 57th Street, New York, NY 10022, USA
296 Beaconsfield Parade, Middle Park, Melbourne 3206, Australia

First published 1985

Printed in Great Britain at the University Press, Cambridge

Library of Congress catalogue card number: 83-24310

British Library cataloguing in publication data

Linggard, R.
Electronic synthesis of speech.

1. Speech synthesis
I. Title
612'.78 TK7882.S65

ISBN 0 521 24469 2

TM

Contents

Preface

This book is intended primarily as a comprehensive text and reference source for scientists and technologists working in the field of speech synthesis. However, the work should be of use as a textbook for courses on speech processing which include a significant component on synthetic speech. It is also hoped that the book is sufficiently clear to enable an intelligent and numerate layman to obtain an understanding of the motives and methods of generating synthetic speech.

In the forty years since 1940, synthetic speech has risen from being a laboratory curiosity to being a commercially viable communication technique. The remarkable success of synthetic speech has only been possible because of contributions made by many different branches of science. The main objective in preparing this text has been to bring together these many disparate theoretical elements into a coherent whole. In particular, an attempt has been made to explain the rudiments of articulatory and acoustic phonetics, acoustic tube theory, digital filtering, transmission line theory, and linear predictive coding. The one component of speech synthesis theory which has not been tackled in any great depth is the phonology and prosody of speech. A detailed exposition of this important topic has not been attempted, partly because it would have lengthened the book inordinately, but mainly because an agreed comprehensive theory does not yet exist and is still the object of considerable research.

A major concern in planning this work has been to give the subject an historical perspective, and to bring out the essential differences and similarities of all speech synthesis devices invented to date. To this end, a computer search of the literature was instituted, which found over six hundred references to 'speech synthesis' since 1968. By back-tracking from these references, a further one hundred or so publications were discovered.

It is hoped therefore that in the research for this book no relevant source has been overlooked. However, for reasons of clarity and brevity, it has not been possible to refer to all the references consulted on any particular topic; the ones cited are those which in my opinion best illustrate the point being made.

R.L. 1983

Acknowledgements

My first thanks must go to the Queen's University of Belfast, for the sabbatical year 1981/2, during which the bulk of this book was written, and in particular to my colleagues in the Department of Electrical and Electronic Engineering, who took over my academic duties during that year.

For the spectograms in chapter 2, I am very grateful to Barry Dupree and John Holmes of the Joint Speech Research Unit, Cheltenham, England, and to the Controller of Her Majesty's Stationery Office (Crown Copyright, 1981) for permission to publish them.

But mainly I must thank the authors of the one hundred and fifty or so references I have cited. The ideas in this book are theirs, and I must particularly thank those from whose work I have taken diagrams and tables. They are acknowledged individually in the figure captions. However, special mention must be made of two sources of these references. The first is Wolfgang von Kempelen, whose book *The mechanism of speech and a description of a talking machine* (published in 1791) has been a constant inspiration to me, and has provided many apt quotations throughout this present work. The other is J.L. Flanagan, whose work I have referred to more than any other, and whose sustained effort over three decades has made significant contributions to every branch of electronic speech synthesis.

Finally, I would like to thank my colleagues and friends whose moral support has sustained me throughout the long year of 1981/2. Not the least I must thank my research students whose enthusiasm and industry is always an inspiration. But in particular I owe thanks to Lore Buchler, Frank Holmes, Ivan Boyd, Rod and Ellen Cowie, Dr E. Ambikairajah and Yiannis Attikiouzel.

1 *History of speech synthesis*

From talking gods to talking computers

'In the beginning was the Word, and the Word was with God, and the Word was God.'
St John 1.1

1.1 Talking gods

The faculty of speech has always been regarded as a god-like attribute, and man has accepted his own ability to speak as clear evidence of his semi-divine nature. It is perhaps for this reason that legends of gods speaking to humans are widely believed or, if not, at least invested with mystical or religious significance, whereas stories of animals conversing with men are generally discredited, and relegated to the status of fairy-tales. No wonder then that mankind's first efforts at speech synthesis were made by priests, who sought to enhance the stature of their gods by giving their images this god-like power. Thus it was found, at the beginning of the Christian era, when many pagan idols were being torn down, that some had been fitted with devices whereby a concealed priest could speak, via a tube, through the mouth of the statue. One in particular, the Head of Orpheus at Lesbos, a famous oracle in Greek times, was found to contain a speaking tube designed to carry a voice from the rear of the oracle to the lips of the head.

The literature of the ancient world is rich in stories of speaking gods and oracles. How much these were due to man's artistry, and how much to man's imagination, it is now impossible to tell. But even in the Christian age talking heads were a recurrent phenomena. Both Pope Sylvester II and Roger Bacon were rumoured to possess 'brazen heads' which had the power of speech. Other heads were reported to be of wood and some even of pottery. Perhaps the best account of such a device, and certainly the most amusing, is given by Cervantes in 1615, in his book *Don Quixote*. He tells how Don Antonio plans an entertainment, in which Don Quixote will be confronted by an 'enchanted head'. This cunning device is described as follows:

> the head, which looked like a bronze bust of a Roman emperor, was all hollow, and so was the top of the table, into which it fitted so neatly, that no sign of a join could be seen. The stand of the table was also hollow, to correspond with the throat and head of the bust, and the whole was made to communicate with another room beneath the one where the head stood. Through all the hollows in the stand, the table, the throat and breast of the bust, ran a metal pipe, so well fitted as to be completely concealed. In the room below, corresponding to the one upstairs, stood the man who was to make the answers, with his mouth to this same pipe, so that the voice from above came down and the voice from below sounded up, clearly and articulately as through an ear trumpet,...
>
> Cervantes, 1970

When the head speaks in response to questions, the gullible Don Quixote is astonished, and believes the head to be enchanted. However, the more down-to-earth Sancho Panza is not so easily impressed, and complains because the head cannot predict the future. 'Animal!' cries Don Quixote, 'What do you expect? Is it not enough that the replies the head gives correspond to the questions asked it?' (These two extremes of attitude are still to be encountered in modern reactions to speaking computers.)

Speaking tubes of the kind described above are more akin to telephones than to speech synthesisers. They do, however, indicate that even in those early times men had the need and the interest, if not the technology, to make synthetic speech. Neither Greek science nor Roman engineering seem to have made much progress in the investigation of speech, synthetic or natural. The world had to wait until the Renaissance before speech became a respectable topic for scientific study. But before following it thither, it is worthwhile considering how nature herself created the problem in the first place.

1.2 The evolution of speech

It is necessary first to explain why man alone, in the whole animal kingdom, possesses the power of speech. Though other animals use cries and calls to communicate with each other, such sounds are instinctive and are far removed from the conscious, careful articulations of human speech. A dog for example, raised in isolation, will bark, growl, whine and yelp in exactly the same way as any other member of its species. These sounds are simply instinctive reactions to fear, anger, hunger and pain, and are no doubt instinctively interpreted as such by other canines. Men also make instinctive noises, such as screams, groans, laughter etc., which have a meaning beyond any particular language or culture. But the formal articulations of speech are precise symbols, which vary from language to language, and which have to be learned – even though the capacity to learn them may be innate.

In discussing the evolution of speech in man, it is necessary to distinguish between physiological evolution, which is transmitted by genetic mutation, and the evolution of technique, which is passed on by learning. The time scales of these two kinds of evolution are also different. The development of technique may take place much more rapidly than physiological evolution, and each of the 4000 or so languages in the world today probably 'evolved' over the last 10 000 years. However, taking one million years as the nominal time span over which the capacity for speech evolved, it seems unlikely that there has been any major change in physiology to cope with either speech production or speech perception.

Certainly, at the peripheral level, the mechanism of hearing in man is not significantly different from that of other higher mammals. In fact, the hearing of the dog is superior to that of man in both sensitivity and frequency range. Likewise, the articulators of speech in humans, apart from their detailed shape, do not differ significantly from those of our nearest evolutionary neighbours, the apes. In fact the main trend in the evolution of head shape seems to have been towards increased cranium size, to accommodate a larger brain. What seems likely, therefore, is that evolution has taken place primarily at the neural level, giving increased processing power. Thus, existing auditory neural networks may have adapted to recognise 'speech-like' sounds, at the same time as muscle coordination sequences have evolved to make such sounds. The dominating factor in this process is likely to have been the faculty of hearing, which, having been perfected over the whole time span of animal evolution, is least capable of being changed in the one million or so years that speech has taken to evolve. On the other hand, given mobile articulators (evolved to manipulate food), the range of vocal noises that can be made is vast, requiring only the enlargement of certain muscle groups and the development of appropriate muscle-coordination sequences. That speech articulations have evolved largely to suit existing perceptual abilities seems more likely when one considers that no less than six of the twelve cranial nerves are involved in speech production.

The essential problem, then, is how the particular set of sounds which constitute speech evolved to be the basic symbols of human communication. The most probable explanation seems to be what Paget called the 'gesture theory'. This supposes that our primitive ancestors once communicated by a system of gestures, both hand signals and facial expressions. Gradually, as hands became involved with manipulation of tools and the mouth became less necessary for gripping and holding, the communicative burden shifted and became concentrated on facial expressions. (Chimpanzees still communicate in this way.) Furthermore, in order to draw attention to the fact that a facial expression was being communicated, primitive man made an instinctive cry, probably a growl. It soon became obvious that, for some facial expressions, it was not necessary to look at the communicator, because the position of the facial muscles were communicated in the 'tone' of the attention-seeking growl. Thus, by a process of learning and evolution, those expressive gestures which gave the clearest auditory indications were favoured as the sounds of speech. And, thus, facial expressions migrated to the mouth, and into the mouth, to become lip and tongue movements – the articulatory gestures of speech.

This theory is supported by modern research (McGurk & MacDonald, 1976), which has shown that visual information (lip movement) is still an important stimulus in the perception of speech. For example, subjects hearing 'ga' and simultaneously seeing 'ba' usually perceive 'da', a fused version of the two syllables, whereas without the visual stimulus 'ga' is perceived correctly. The illusion that 'da' is being heard is very powerful, and prior knowledge does not interfere with the percept. This indicated that the integration of the aural and visual information is taking place at an unconscious level. It would appear then that speech sounds are simply clues to the posture of the speech articulators. The fact that visual evidence can override the aural seems to indicate that such postures may have derived from visible, facial expressions.

At higher levels, too, there seem to be indications that language ability in man is unique. Chimpanzees cannot, of course, be taught to speak, but they can be trained to use sign language, which they learn with facility. However, having acquired a basic vocabulary, they rarely go beyond the simple naming of objects and actions. They appear to lack the ability to learn the more complex processes of language. In fact, when one considers how intimately thought and language are interconnected, it seems even more unlikely that chimpanzees should learn human language, since this would imply a capacity for human thought.

1.3 Von Kempelen

The first major effort in experimental speech research was made by Wolfgang von Kempelen, a Hungarian nobleman in the Austrian Imperial

Service. Von Kempelen was a talented engineer with a genius for mechanism. In 1769 he invented a chess-playing 'automaton', which became famous throughout Europe and is said to have played and beaten Napoleon. This machine, consisting of a large cabinet on top of which stood the chess board, was in fact an elaborate illusion, the essential part of its 'mechanism' being a concealed chess player. The machine was exhibited in various parts of Europe and the United States, and owed its success to the fact that the cabinet was so cleverly designed that it always appeared to be empty.

Von Kempelen conceived and constructed his 'chess-man' in six months and, pleased with his success, he turned to what appeared to be a much simpler problem, the construction of a talking machine. This task was to become his absorbing interest for the next twenty years, almost the rest of his life. An account of how he arrived at his final design, and the discoveries he made concerning the nature of speech, are given in his book *The mechanism of speech and the construction of a speaking machine*, published simultaneously in German and French in Vienna in 1791. This great work, which shows considerable practical insight into the nature of speech, is also very nicely written, and provides many apposite quotations.

Von Kempelen's first concern in designing his talking machine was to find a suitable voicing source, to imitate the action of the human vocal cords or glottis. After considering and rejecting many musical instruments, he finally seized upon the drone reed of the 'humble bagpipe', which he came across by chance in a country inn. Experimenting with a kitchen bellows (for lungs), the reed (as glottis) and the bell-end of a clarionet (as mouth), he managed, by using his hand, partially to obstruct the air flow, to produce a few vowel-like sounds. However, after this initial success further progress became much more difficult. He noted that 'in order to continue my experiments it was necessary, above all, that I should have a perfect knowledge of what I wanted to imitate. I had to make a formal study of speech and continually consult nature as I conducted my experiments. In this way my talking machine and my theory concerning speech made equal progress, the one serving as guide to the other.'

Like so many before him (and even some after), von Kempelen at first misunderstood the nature of vowels. He admits that 'it seemed to me that the vowel "i" was much higher in tone than "o".' He was confusing pitch, the fundamental frequency of vibration of the vocal cords, with formants, the resonant frequencies of the vocal tract. Later, experimenting with an organ, he observed that 'every pipe I sounded, large or small, produced "a" only: thus "a" was sharper or flatter (higher or lower) according to the size of the pipe, but it remained, nevertheless still only "a" '.

Once he had realised that pitch does not contribute to vowel quality, von Kempelen simplified his experiments by tuning all the organ reeds to the same note. He then set about modifying the nature of these sounds by means of various obstructions in the pipes. Soon he had obtained several vowels, and began to think of making some consonants. Again progress was slow, and it took him two years to discover how to make the sounds 'p', 'm' and 'l'. He was now in a position to synthesise syllables or even whole words. He notes, 'Each letter had its own key, causing it to sound when it was pressed. But what happened when I wanted to combine some of them to form a syllable or a word? Two very disagreeable things....' These 'disagreeable things' will be familiar to modern speech researchers, as the problems of co-articulation, the phenomenon whereby speech sounds merge into and overlap each other.

Eventually, von Kempelen realised that 'it was possible, following the methods I'd been using, to invent separate letters, but never to combine them to form syllables, and that it was absolutely necessary to follow nature which has only one glottis and one mouth, through which every sound emerges and which gives a unity to them'. Thus he began again, bravely abandoning the various lines of research he had been pursuing for many years. He comments, 'Suffice it to say that a powerful horse would have had difficulty in pulling a cart loaded with the designs I had to reject.'

Finally, after many more years of patient research, von Kempelen arrived at the machine shown in Fig. 1.1, which could speak whole phrases in

Fig. 1.1 The talking machine of Wolfgang von Kempelen.

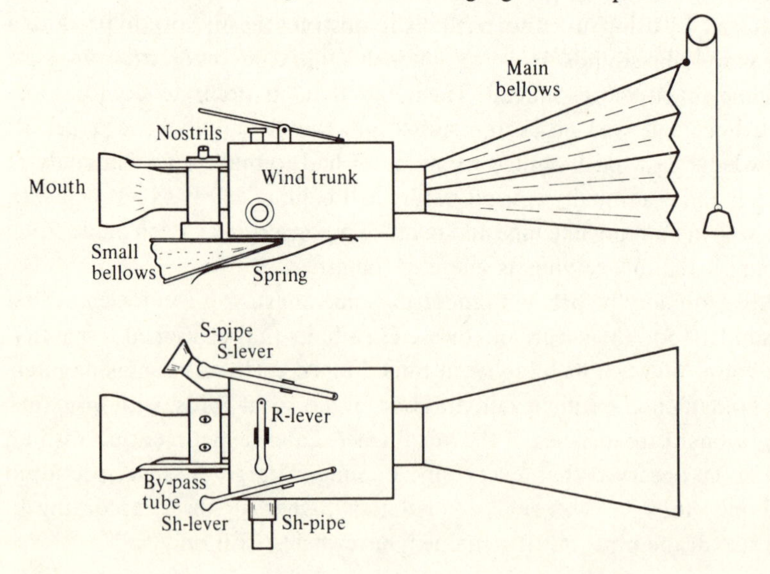

French and Italian. The main part of this talking machine is a small box measuring three and a half by two and a half by one and a half inches. This 'wind trunk' is fed with pressurised air from a large bellows. Its main outlet is covered by the reed assembly, which consists of a sliver of wood held over the aperture by a brass wire. With sufficient air pressure, the reed vibrates on a monotone. A lever on top of the wind trunk can cause a rod to interfere with the reed's vibration, to give an 'r' sound. Two other levers on top of the box permit air to escape through special pipes, which give the 's' and 'sh' fricative sounds.

The air which vibrates the reed passes out through a 'throat' to a 'mouth', as shown in Fig. 1.2. Let into the top of the throat are 'nostrils', consisting of two short pipes, and attached to the bottom is a small bellows. This bellows, compressed by a spring, simulates the expansion of the vocal tract in humans, by allowing air to expand into it against the spring. Thus air can continue to flow through the reed after the mouth opening has been closed. This is an important feature of voiced stops, 'b' for example. Also, it provides a reservoir of compressed air for the simulation of plosive sounds, such as 'p'. The mouth is a cup, made of india-rubber so that it can be easily shaped to give the various vowel sounds. The mouth also receives air directly from the wind trunk, via a by-pass tube, to provide pressure for unvoiced sounds. After making a special pipe for the 'f' fricative, von Kempelen dispensed with it when he found that air, escaping through holes made for the lever mechanism, gave the same noise.

The machine is 'played' in the following manner. The right elbow is used to compress the main bellows, which is normally held open, while the right hand lies along the top of the wind trunk. The fingers of the right hand are used to close the mouth and the nostrils, and to operate the three levers. The left hand is used exclusively to manipulate the 'lip' of the mouth, to produce

Fig. 1.2. Details of the 'mouth, throat and nostrils' of von Kempelen's talking machine.

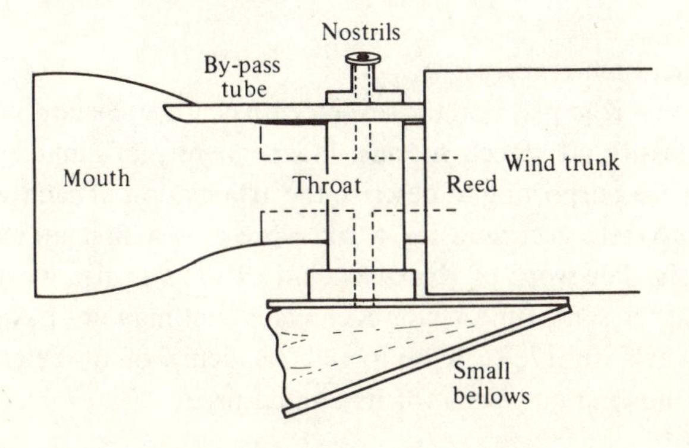

voiced sounds. With the bellows depressed, the nostrils covered, and the mouth open, the reed vibrates and the left hand is able to make all the vowels, as well as 'l', 'w' and 'y' sounds, by changes of position. With the mouth closed, the sounds 'n' and 'm' are obtained by uncovering one or both nostrils. Plosives are made by closing the mouth and releasing the pressure suddenly. For 's', 'sh' and 'r', the appropriate levers on top of the box are depressed. The 'f' fricative is made by closing the mouth and the nostrils, and increasing the pressure so that air escapes noisily, through the leaks. Voiced versions of all fricatives and plosives can be produced by allowing air simultaneously to vibrate the reed, and to pass out through the mouth.

Von Kempelen admits that the machine cannot reproduce 't', 'k', 'd' and 'g', because it lacks a proper tongue. After giving his ideas on how this deficiency might be overcome, he confesses that in practice he uses a 'p' for 't' and 'k', and a 'b' for 'd' and 'g'. He notes, candidly, that listeners are easily deceived. 'One is especially easily mistaken when one imagines in advance the words that the machine must say, and when it pronounces that word one imagines it has been heard correctly. And if it should happen that a practised ear notices the defect, there is always a great advantage in the childish tone that the machine possesses. One easily forgives a child, who stammers occasionally, for using one letter instead of another: it is enough to have understood what he meant.'

In an era of gentleman-scientists, von Kempelen was more of a aristocrat-engineer. His main objective was not to produce an elegant theory, but to construct a practical, working machine. This heavy constraint gives a seriousness and directness to his work which is, in many ways, prophetic of modern technological methods. There can be no doubt that von Kempelen understood many of the important mechanisms of speech production, and the success of his talking machine confirms this. For those who followed him, he was not only a source of practical information, but also an example of persistence and ingenuity.

1.4 The nineteenth century

Before von Kempelen, in the seventeenth century, Bishop Wilkins attempted to classify all speech sounds, independently of language. He produced diagrams purporting to describe the articulation of each sound. Whilst not completely accurate, his work represents a first attempt at scientific analysis. The work of Kratzenstein in Russia is also worthy of note. He lived at the same time as von Kempelen, but may not have been aware of his work. In 1769, the Imperial Academy of St Petersburg proposed the following questions for its annual prize:

1. What is the nature and character of the sounds of the vowels a, e, i, o, u, that makes them differ from one another?

2. Can an instrument be constructed like the *vox humana* stops of an organ which accurately expresses the sounds of these vowels?

The prize was won by Professor Kratzenstein, who constructed five resonant cavities which gave the required sounds when excited by a reed. The shapes of these cavities, illustrated in Fig. 1.3 were said to have been based on measurements taken from a real vocal tract, though the resemblance may seem somewhat obscure to modern eyes. However, the reed used by Kratzenstein was an innovation, being a 'free' reed (like that of a mouth organ), in contrast to the 'beating' (bagpipe) reed of von Kempelen.

The next important investigation of vowel sounds was by Robert Willis who, in 1829, experimented with uniform tubes. Willis believed that the quality characteristic of vowel sounds was due to a single resonance of the vocal tract, and that this resonance was the same for men, women and children. Willis measured these resonant frequencies from his tube model for a number of vowels and some actually do correspond to what are now known as 'formant' frequencies. Wheatstone, in about 1837, constructed an improved version of von Kempelen's speaking machine. Experiments based on this apparatus led him to the conclusion that vowel quality was due to multiple resonances.

The problem of the exact nature of vowel quality was also studied by the great German physicist Helmholtz, famous for his analysis of spherical resonators. His opinion was that the back vowels, like 'o' and 'u', were due to single resonance, and front vowels such as 'e' and 'i', had double resonances. He apparently deduced this from the relevant vocal tract shapes, on the assumption that double resonance must be due to a double cavity. The idea that each vocal tract resonance must have its own particular cavity, was to confuse research on vowels even into the twentieth century.

In 1873, Potter made artificial vowels using an india-rubber sphere excited by a reed. By compressing the sphere into different shapes, he was able to imitate many of the vowel sounds. But he seems not to have made

Fig. 1.3 Kratzenstein's resonators.

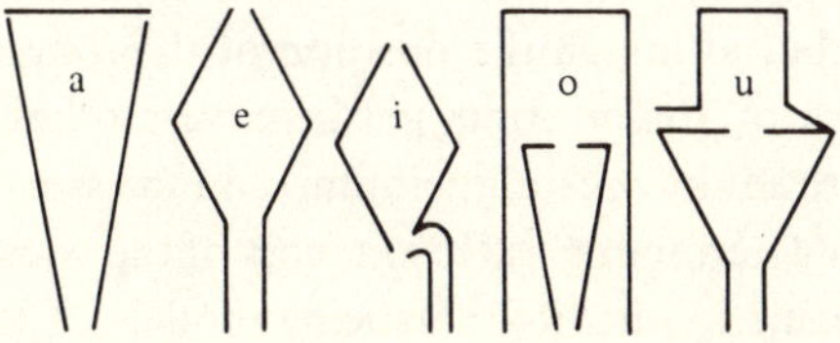

a e i o u

any measurements of the relevant shapes or their resonant frequencies. About the same time, in the USA, Graham Bell was making his investigation into vowel sounds. His conclusion was that vowel quality was due to double resonance. R.J. Lloyd, in about 1896, actually measured the resonances of the vocal tract for various articulations. His table of resonant frequencies for the vowels is surprisingly accurate, if incomplete. However, his main contribution was his realisation that vowel quality is due, not to the absolute value of these resonant frequencies, but to the relative intervals between them.

By the end of the nineteenth century, instrumental methods of acoustic measurement were being widely applied. In 1916, D.C. Miller actually succeeded in making a mechanical Fourier analysis of vowel waveforms. He managed to resynthesise the vowel sounds from their harmonic series, by using a set of organ pipes, each corresponding to a particular Fourier component. The use of the oscilloscope in the 1930s and the sound spectrograph in the 1940s made possible the investigation of speech in detail never before imagined. But, before discussing this, it is worthwhile pausing to consider the brilliant, intuitive work of Sir Richard Paget.

1.5 Sir Richard Paget

Paget (1930) first became interested in speech after hearing words intoned without accompaniment, in the 'plainchant' of English church music. He was puzzled by the way in which different partials (harmonics of pitch) were emphasised by different vowels. In 1916, he began an informal study of these effects, using as his measuring instrument his own, particularly fine, musical ear. Paget trained himself to hear the details of speech sounds, ignoring the fact that the sounds had meaning. Thus he was able to identify individual harmonics in voiced speech, and the resonances in whispered vowels. By these means he constructed a vowel resonance chart, showing the two main resonances for all the vowels, and for some, an 'additional' resonance. This is given in Fig. 1.4, and some modern vowel formant frequencies are superimposed for comparison. It can be seen that the resonance ranges given by Paget are quite accurate, even for the third formant, Paget's elusive additional resonance.

Paget taught himself to manipulate the resonant frequencies of his own vocal tract, so that he could vary the first and second formants almost independently. He noted that although the position of the tongue-hump did have a significant effect on these formants, there were other factors. Both the size of the aperture left by the tongue-hump and the size of the lip opening were significant in determining the resonances of the vocal cavity. He proceeded to confirm this by making plasticine models of the vocal tract, which amply demonstrated his double-resonance theory. This study developed into a general investigation of resonant cavities. He showed how cavities could be connected in series or in parallel, and noted that the parallel connection was to be preferred because it permitted independent tuning. As well as the vowels, he also synthesised the consonants p, t, k, r, f, v, m and n, all using resonant cavities. Before an 'audience interested in phonetics', he performed a recognition test, which must have been the first psycho-acoustic experiment involving synthetic speech. All the sounds were correctly identified by a majority of his listeners.

In a later study of French vowels, Paget was able to identify the extra, nasal resonance, and correctly attribute it to air resonating in the nasal cavity. He also observed that nasal 'twang' was independent of the nasal

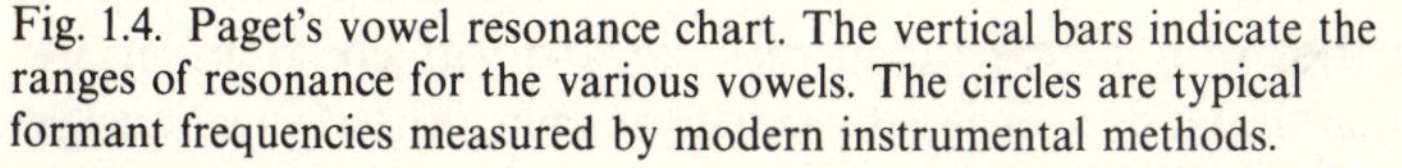

Fig. 1.4. Paget's vowel resonance chart. The vertical bars indicate the ranges of resonance for the various vowels. The circles are typical formant frequencies measured by modern instrumental methods.

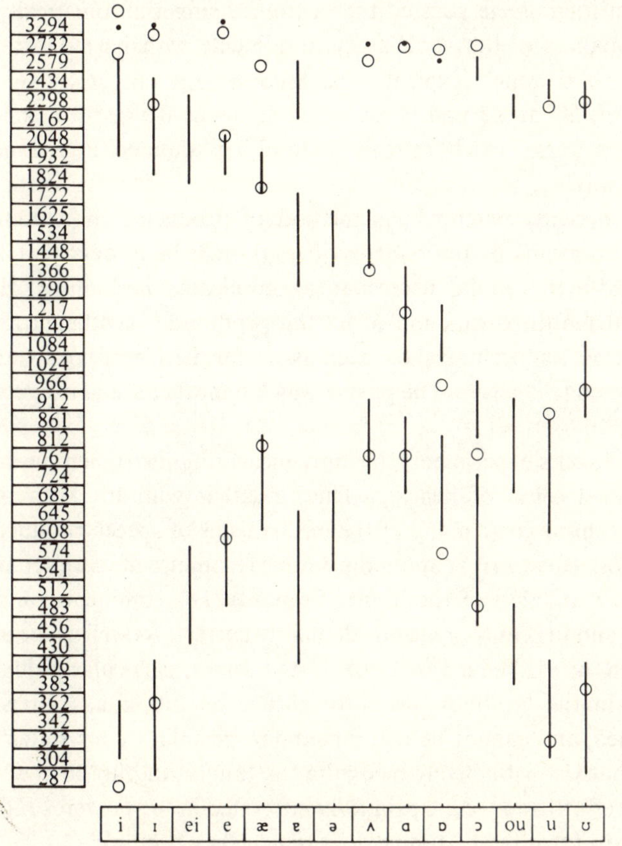

cavity, and concluded that it was caused by lowering of the velum, which distorted the normal resonances. In an analysis of consonants in real speech, Paget was able to detect the transient resonances of plosives. His remarkable hearing enabled him to observe that these resonant frequencies varied, depending on the vowel with which the plosive was associated. His conclusion that 'the different plosives were presumably due to the presence of the different initial resonances – at the moment of the release of the closed orifice that produced them, or to the different changes which the initial resonance underwent as the closure became fully released, or to both of these causes', is absolutely correct. Paget also produced resonance charts for plosives and fricatives.

In his study of vocal cord vibration, Paget experimented with a human larynx, excised from a cadaver, complete with tongue. Using a compressed air supply to excite the cords, he noted that the frequency of vibration varied with increase of air pressure, and with cord tension. He observed that the voice pitch is substantially independent of vocal tract resonances, except in singing. He guessed that a trained singer automatically adjusts the resonances of the vocal tract to coincide with harmonics of pitch, to give good 'tone'. Typically, he followed up this study by making models of the larynx and vocal cords. Later he devised a novel kind of motor-car horn, which instead of honking, shouted in a loud voice, 'away! away!'

Paget actually patented his method of producing individual speech sounds by means of resonant cavities, though he admits that to make connected speech would 'necessitate a complicated mechanism'. However, he does have the distinction of having produced a synthetic phrase, by 'conducting' an 'orchestra' of seven assistants, each equipped with one of his patented resonators. The phrase was 'oh mother are you sure you love me'.

In all Paget's experiments, his only measuring instrument was his own finely-tuned sense of hearing. This, together with his keen intuition, enabled him to guess many of the mechanisms of speech production and perception. He was right about the double resonance of vowels, but did not realise the function of the higher formants. His comments on nasality, voicing, and the consonants are all mainly correct. As well as the acoustics of speech, he studied a variety of related topics, particularly singing. His interest in the 'gesture' theory of speech evolution has already been mentioned, and cannot be better summarised than by his statement 'the fact is that we instinctively recognise the tongue and lip posture by their acoustic effect, and we are primarily interested in the postures rather than in the waveform or the tone colour which they produce'.

1.6 Electronic synthesisers

After 1930, the advent of electronic instrumentation, in particular the oscilloscope and the sound spectrograph, opened the way to a complete understanding of speech acoustics. Furthermore, artificial speech could now be synthesised using electrical circuits, and there was only one more attempt to make a mechanical talker. This is the machine of Riesz, which, though mechanically very sophisticated, could not compete with the electrical synthesiser's superior flexibility. Stewart (1922) was the first to make electrical vowel sounds, using an electro-mechanical interrupter to excite a doubly-resonant electrical circuit. But the first, truly electronic speech was synthesised by Dudley (1936), in his invention, the vocoder. This device analyses speech into low bandwidth spectral components, and recombines them to reproduce the original speech. The purpose of the vocoder is to reduce the transmission bandwidth of speech, in a technique known as analysis–synthesis telephony, and it was principally for this application that interest in speech synthesis was revived.

Dudley's vocoder used a ten channel filter bank to provide a spectral analysis of the speech, and a special circuit to detect the pitch of the voicing. The speech was resynthesised by using pulses at pitch frequency to excite a set of ten filters, the gains of which were proportional to the values of the ten spectral components. The synthesis half of this machine was later modified to permit manual control, so that it could be 'played' via a system of keys and switches. It was, in effect, a realisation of von Kempelen's dream of a 'speaking organ'. This machine was demonstrated at the New York world fair, in 1939, where trained operators were able to produce synthetic speech to order (Dudley, Riesz & Watkins, 1939).

Other electronic synthesisers soon followed, all built as part of vocoder research. Controlling such machines by hand-operated keys was not really very satisfactory, because it made the accurate specification of control parameters virtually impossible. A popular alternative was to paint the control functions onto transparent material and use optical scanning to convert these to electrical signals. The machine of Lawrence (1953), and the pattern play-back devices of the Haskins laboratory (Cooper, Liberman & Borst, 1951), are typical examples. From the mid-1950s onwards, the availability of digital computers provided a more systematic way of supplying electronic synthesisers with control parameters. Computers also made it possible to simulate speech synthesisers by program. Though this could not be done in real-time, it did enable research in speech synthesis to be carried on without complicated hardware synthesisers. It also allowed the performance of any proposed synthesiser to be evaluated and optimised without its actually having to be built.

By 1960, two different approaches to the design of speech synthesisers had become manifest. In one, the speech production mechanism itself is modelled (as an electrical or computational analogue), with more or less physiological detail. In the other, it is the speech signal that is modelled, using whatever means are convenient. This latter method leads to what might be called a 'signal model', in contrast to the 'system model' of the former approach. This dichotomy still persists, and to a certain extent reflects the division of interest, between those who seek a deeper understanding of the speech production mechanism and those intent on synthesising speech by the most economic and direct means. Not surprisingly, signal models are easier to design and construct, since it is only necessary to model the effects of articulation, rather than its mechanism. On the other hand, system modelling, or articulatory synthesis, as it is usually known, is very difficult to achieve, and the most complete models are computational analogues, which run as programs on digital computers.

The advent of digital computers had a very profound effect on speech research, besides that of making it easier to design, implement and control synthesisers. The possibility of using synthetic speech as an output medium for digital computers gave a new direction and motivation to the field. By about 1970, speech synthesis for computer output had become a more important stimulus for research than vocoder design. In fact, about this time interest in vocoders began to fall off, for two reasons. The first was that new technology had made bandwidth in telecommunication systems both cheap and plentiful. The second reason was that, for most purposes, the vocoder problem had been solved anyway, by the use of Linear Predictive Coding (LPC). LPC is an analytic method of encoding speech as low bandwidth parameters, and of resynthesising these with virtually no loss of information. The LPC algorithm makes it possible to encode speech for transmission (or storage), at data rates as low as 1000 bits per second, a huge saving over the 64 000 b/s of conventional Pulse Code Modulation (PCM). The LPC algorithm has been used extensively to parameterise speech for use as voice output from computers and other electronic machines. In this application, standard words and phrases are parameterised, and stored as a vocabulary. They are then accessed as required by the machine to provide voice output. This is not a very satisfactory way of synthesising speech, since, for it to be in any way flexible, a very large vocabulary is required. Also, for long utterances, it is necessary to modify the form of the word parameters to suit the particular context of each word. Fortunately, more efficient means of synthesising speech, using phonetic specification, became available, again mainly through the application of digital computers.

The use of vocal tract analogues of various kinds has advanced the understanding of speech production mechanisms to a considerable extent. In particular, Fant (1960) made a study of vowels, and showed that there are not just two or three vocal tract resonances, but an infinite number. (He thus refutes the theory that each resonance must be associated with a particular cavity.) However, in the frequency range of interest, upto 5 kHz, there are only four or five formants. In general terms, two of these are necessary to specify vowel quality, a third is required to establish speaker identity, and the fourth/fifth may be added to give natural voice quality. Other models (Coker, 1976) have been used to relate formant frequencies to positions of the articulators. Flanagan, Ishizaka & Shipley (1975), using a particularly detailed model of the vocal tract, have shown how fricatives are generated by turbulent air flow. Vocal cord models have largely confirmed Paget's speculation that cord vibration is controlled by lung pressure and cord tension, and that pitch is independent of vocal tract resonances.

The early speech synthesisers were all constructed using analogue circuitry. Those of the signal model kind (for example Flanagan, 1957; or Holmes, Mattingly & Shearme, 1964) tended to use three or four resonant circuits to simulate the effect of formants. The articulatory or system models (for example Dunn, 1950; or Stevens, Kasowski & Fant, 1953), used discrete inductor–capacitor transmission lines to model the vocal tract. In the interests of reliability and accuracy, analogue designs were superseded by either digital circuits or computer simulations (Rabiner, 1968*a*). Essentially, the more sophisticated, system, models (articulatory) are simulated on digital computers, whereas the simpler, signal, models are realised as digital circuits. In particular, the extreme simplicity of the LPC synthesiser has made it possible for them to be constructed as single-chip integrated circuits.

Digital computers have been used not only to supply speech synthesisers with suitable control signals, but also to generate them. Control parameters can be extracted from real speech by use of a suitable analysis technique, such as the LPC algorithm. Alternatively, they can be generated from a phonetic specification, using a system of rules which take into account the acoustic effects of each phoneme. This 'synthesis-by-rule' (Holmes *et al.*, 1964) is a much more flexible technique for synthesising speech than parameterising human utterances. It permits virtually unlimited vocabulary, and allows the effects of word context to be included in the rule system. The main difficulty in rule synthesis is that the original phonetic specification must be written by a skilled phonetician. However, this problem can be overcome (Teranishi & Umeda, 1968) by using an additional set of rules to obtain a phonetic specification directly from ordinary English text. Such 'synthesis-from-text' algorithms have become extremely sophisticated

(Allen, 1976), and they embody a considerable amount of knowledge about the grammar and syntax of the language. The information used in these rule based synthesis programs has been gathered from experiments with synthetic and natural speech, designed specifically to reveal the systematic structure of speech at the phonological and linguistic levels (for example Umeda, 1975; and Klatt, 1976*b*).

Just as it was impossible to discuss the evolution of speech production without considering speech perception, so it is difficult to talk about the history of speech synthesis without mentioning speech recognition. The vocoder is an analysis–synthesis machine, and both machine recognition and synthesis of speech have their roots in this invention. However, the progress of speech recognition has not been as spectacular as that of synthesis. The probable reason for this is that human speech perception relies on the superb pattern recognition abilities of the human brain, about which very little is known, whereas the details of speech production, even if not understood, can be copied quite faithfully. The speech recognition machines available at present are essentially spectral pattern recognisers. They are particularly effective in recognising isolated words from a speaker for whose voice they have been trained. They cannot, however, recognise individual phonemes in continuous utterances, even for a single speaker. Though they can identify spectral patterns with considerable facility, they are unable to convert these to the 'tongue and lip postures', which appear to be the invariants of human speech. This limitation in speech recognition also places a limit on the degree of perfection to which speech synthesis may aspire. This is because, at the higher levels of the speech production mechanism, there is a component of feedback from the speaker's perception of his own speech. The present generation of speech synthesisers may be likened therefore to deaf speakers, whose speech is not self-monitored, and who rely on their memories of past speech to provide current production information.

1.7 The future

Since the speaking Head of Orpheus correctly prophesied the death of Cyrus the Great in the sixth century BC, speech synthesis, as a science, has made considerable advances. Though most of this progress has been in the last fifty years, it must not be forgotten that the notions of speech production current in the 1930s were the hard-won results of experimenters like Paget and von Kempelen. Modern parametric speech is virtually indistinguishable from natural utterances, and the latest rule-synthesised speech can be perfectly intelligible, if not very natural-sounding. The future of synthetic speech, therefore, seems assured, particularly as an output medium for our contempory oracle – the digital computer. It is likely that this application will become more important as the performance of speech recognition approaches that of human abilities. It should then be possible to use simultaneous synthesis–analysis of speech, to simulate the feedback effect in speech production. This should go a long way to eliminating the 'machine accent' of rule-synthesised speech. The perfection of speech recognisers will also make possible computers with which (or whom) one will be able to hold a conversation. But, before this millennium, it will be necessary to find out a lot more about the structure and generation of language, so that the computer itself can 'think' what it will say, rather than just reading prepared statements. That this will one day be possible cannot be doubted; however, to do it will probably require nothing less than the complete simulation of the human thought process.

2 Human and synthetic speech: the problem

Articulatory and acoustic descriptions of speech

'Thus in order to continue my experiments it was necessary, above all, that I should have a perfect knowledge of what I wanted to imitate. I had to make a formal study of speech and continually consult nature as I conducted my experiments. In this way my talking machine and my theory concerning speech made equal progress, the one serving as guide to the other.'
Von Kempelen, 1791

2.1 Human speech production

The mechanism of human speech production at the physiological–acoustic level is quite well understood. Unfortunately, the same cannot be said of the psychological and neural levels which immediately precede and control the movements of the speech articulators, that is, the tongue, jaw, lips and velum. It would appear (Perkell, 1981) that sensory feedback plays a large part in this control process. Our vocalisations are continuously self-monitored, partly by aural feedback, and partly by proprioceptive (sense of touch) feedback. The vocal tract itself, particularly the tongue, is rich in sensory nerve endings. Their importance in speech production can readily be appreciated by those who have received local anaesthetic in the course of dental treatment. If any part of the tongue or mouth is numb, it is difficult to speak properly. This, of course, soon wears off, but it is known (Cowie & Douglas-Cowie, 1983) that the degradation of speech facility which ensues after the onset of deafness is permanent and progressive. Thus, it appears that both kinds of feedback are necessary for normal speech production.

It is impossible to characterise the dynamics of any system without a complete knowledge of the feedback paths. Not nearly enough is known about these feedback processes, which are important for both the speech scientist and the engineer involved in the design of speech synthesisers. The problem is even more difficult when the system is non-linear, as it almost certainly is in the case of speech production. Thus, though speech dynamics can be measured and imitated, often quite successfully, they are still inadequately understood.

Any account of speech production is necessarily complicated by the fact that speech can be described on three different levels, the acoustic, the articulatory and the linguistic. There is also a problem of terminology; at the linguistic level, the basic speech unit is called the 'phoneme', which is translated via an articulatory 'gesture' into a 'phone' at the acoustic level. Essentially, the phoneme, a unit of language, is what the speaker intends to communicate, and the phone is the sound he actually produces. Speakers of a given language all use a common set of phonemes; only thus is communication possible. However, at the acoustic level the sounds uttered by different speakers communicating the same string of phonemes will be highly idiosyncratic. Differences at the acoustic level are due to dialect, vocal tract size and shape, and speaker mannerisms. These differences may be considerable, so much so that it is not practical (or possible) for the phonetician to transcribe the actual sounds he hears; instead he attempts to characterise speech in terms of articulatory gestures. At this level, differences due to individual vocal tract size and shape are eliminated, making the problem of representing speech by a set of formal symbols much more tractable. The first description of speech sounds given here is therefore in terms of articulatory gestures, and uses the symbols of the International Phonetic Association (IPA). However, an alternative description of speech is given at the acoustic level, using speech spectrograms to illustrate the speech spectra of a typical male speaker.

2.2 The articulators

Fig. 2.1 illustrates a cross-section (mid-sagittal plane) through the human head, delineating the vocal tract and showing the speech articulators. The vocal tract may be said to begin properly at the larynx (vocal cords), even though the trachea (windpipe) and lungs are obviously involved in vocalisation. The acoustic coupling between the vocal tract and the trachea is small, and has little direct effect on the speech sounds. The lungs do, of course, provide the primary acoustic energy for speech, and it is air forced through the trachea, via the glottis, into the vocal tract, upon which the vibratory patterns of speech are imposed.

There are two principal ways of superimposing vibrations on this air stream. The simplest is to create turbulence by making a constriction at some point in the vocal tract. This effectively excites the tract at the point of constriction with random acoustic noise. The more efficient method of excitation is to cause the vocal cords to vibrate in a periodic manner. These 'cords' are, in fact, curtains of tissue which can be held, with more or less muscular tension, to block the air flow from the trachea. If this tension is not too great the air pressure in the glottis forces the cords apart and a puff of air is released into the vocal tract. This sudden release of pressure in the glottis permits the cords to close again, so that the pressure builds up to another release. By varying the tension of the cords and the air pressure from the lungs, humans can control the frequency of these vocal cord vibrations over a typical range 50–500 Hz for adults, and up to 1000 Hz for children.

The vibrations of vocal cords are similar to the lip vibrations of a musician in playing a brass instrument. In playing a particular note, a trombonist, for example, holds his lips in tension in a similar way to vocal cords and forces air through them. There is, however, one important difference; to change the note, the trombone player must not only change his lip tension, but must also alter the resonant frequency of the instrument. In contrast, the frequency of vibration of the vocal cords is almost entirely determined by the mass–tension system of the cords themselves, and is relatively independent of the resonant frequencies of the vocal tract. The exact degree of interaction between the cord system and the vocal tract is discussed in chapter 3; the main point here is that the coupling is slight enough to enable the excitation source to be considered separately from the resonant system. This source/system separation is fundamental to the design of almost all speech synthesisers.

The vibratory patterns superimposed on the airflow through the vocal tract are said to be 'voiced' (periodic vibration of vocal cords), 'unvoiced'

Fig. 2.1. Cross-section through the human head, showing the speech articulators.

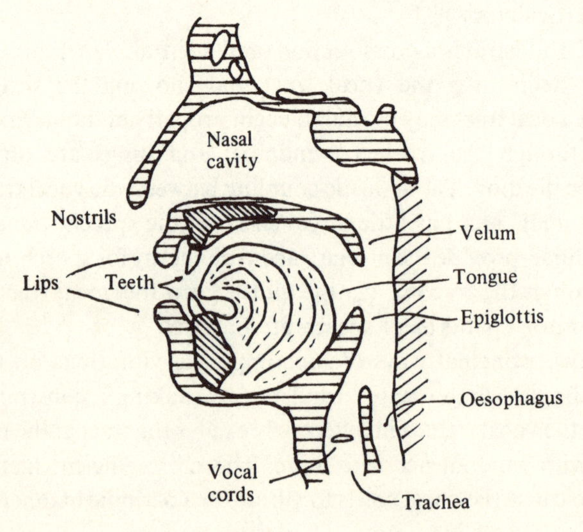

(turbulent flow at constriction), or mixed (periodic and turbulent). Figs. 2.2 and 2.3 show these basic types of excitation as time waveforms. The glottal vibrations of voiced sound (Fig. 2.2) are considerably modified by the acoustic resonances of the vocal tract before they emerge from the lips as speech sounds. Fig. 2.3 shows the waveform of typical voiced and unvoiced speech.

The resonant properties of the vocal tract will depend on the dimensions and detailed shape of the tract at any particular moment in time. By use of the articulators to vary the shape of the vocal tract, the vibratory patterns impressed on the airflow are changed. This is the essential articulatory mechanism of speech.

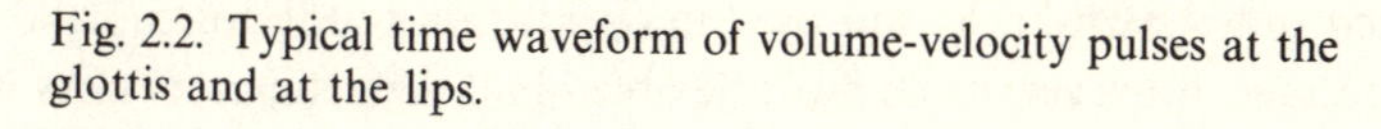
Fig. 2.2. Typical time waveform of volume-velocity pulses at the glottis and at the lips.

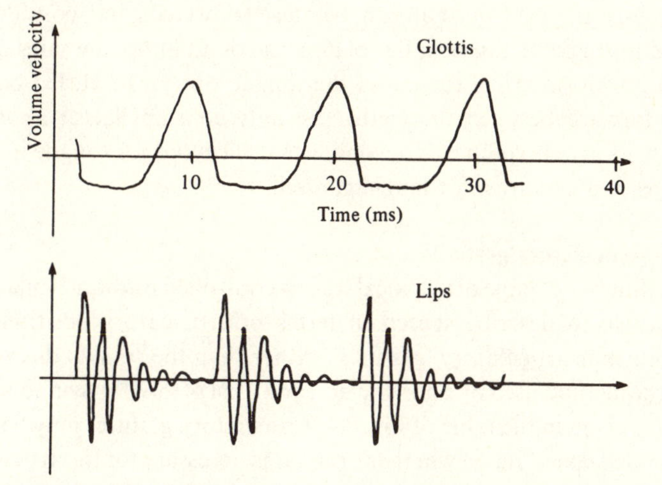

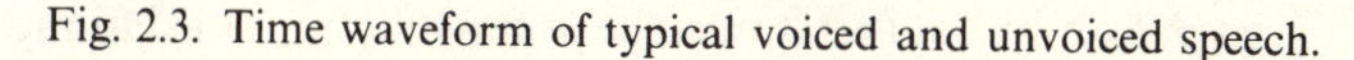
Fig. 2.3. Time waveform of typical voiced and unvoiced speech.

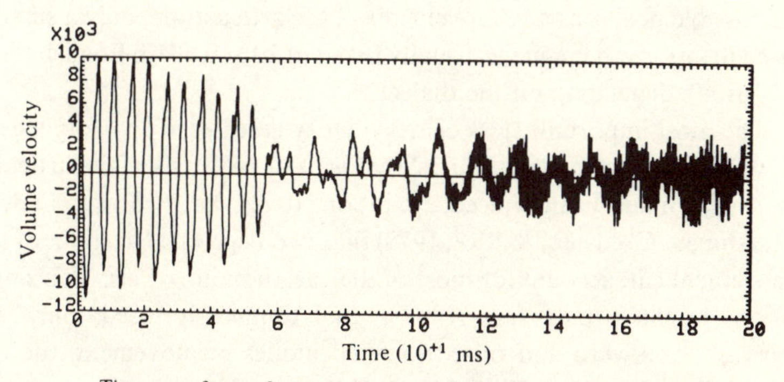

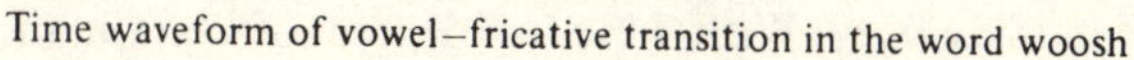
Time waveform of vowel–fricative transition in the word woosh

The principal agents involved in changing the shape of the vocal tract are the tongue, the lips, the jaw and the velum. These are known as the 'movable articulators'. Those parts of the vocal tract against which these movable articulators may be opposed are known as the 'fixed articulators'. When air has passed from the glottis through the pharynx, it may be diverted at the velum, partly or wholly, through the nasal cavity and out through the nostrils. This makes possible the class of speech sounds known as 'nasals'. Since the shape of the nasal cavity is fixed, the only articulation involved in nasality is the raising/lowering of the velum. In contrast to the nasal cavity, the oral cavity is amazingly variable in shape. The jaw can be open/closed to a varying degree, and to a lesser extent it can be moved forward/backward. The lips are even more mobile, and can be open/closed, pushed-out/drawn-back, and used in opposition to the tongue and teeth. The tongue, however, it the most flexible of all the articulators. As well as moving forward/backward it can be raised/lowered, and considerably changed in shape. It can be a flat plate or a round lump, the sides can be folded up to form a half tube, and the tongue tip can be curled back or pushed forward between the teeth. The only constant factor about the tongue is its actual volume. No wonder that the word for tongue in many languages is a synonym for language itself.

2.3 Articulatory gestures

Since the shape of the vocal tract is controlled by the articulators, it is preferable to describe speech in terms of articulatory positions and movements, or articulatory 'gestures', rather than the sounds themselves. This is the method used by phoneticians, and a list of the IPA symbols of RP English is given in the table of Fig. 2.4. Articulatory gestures may be static or dynamic depending on whether or not it is necessary for the articulators to move. A description of the gestures appropriate to the English language will be given presently; but first it must be noted that from the infinite range of possible positions and movements of the articulators, only a small set is used in any given language, usually between 20 and 60. In English there are about 40, depending on the dialect.

The most important class of articulatory gestures in any language is the vowels. These are static gestures, that is fixed positions of the articulators, normally made using voiced excitation. It has been shown (Ladefoged, Harshman, Goldstein & Rice, 1978) that two degrees of freedom of tongue movement can account for most of the variation in vowels. This simplification implies that, for vowels, the tongue is used only in its forward/backward and raised/lowered modes of movement, the tongue shape remaining fixed. With these two degrees of freedom the vowels can be

Fig. 2.4. Some suggested IPA symbols for the phonetic transcription of English.

Vowels		Diphthongs	
i—beat	ʌ—hut	eɪ—rate	ɒə—boar
ɪ—bit	ɒ—hot	aɪ—rite	ʊə—boor
e—bet	ɔ—horn	ɒɪ—lois	əʊ—load
a—bat	ʊ—hood	ʊɪ—louis	aʊ—loud
ə—further	u—loot	ɪə—peer	
ɑ—father		eə—pair	

Stops			
p—pea	b—by	m—ram	*Glottal stop*
t—tea	d—die	n—ran	
k—key	g—guy	ŋ—rang	?—butter

Fricativcs		
θ—thigh	ð—father	*Glottal fricative*
f—fie	v—carver	
s—sigh	z—mars	
ʃ—shy	ʒ—largess	h—hot

Liquids	Semi-vowels
r—rye	w—we
l—lie	j—ye

Fig. 2.5. Vowel plane with front and back raising/lowering of the tongue as axes. (After Ladefoged *et al.*, 1978).

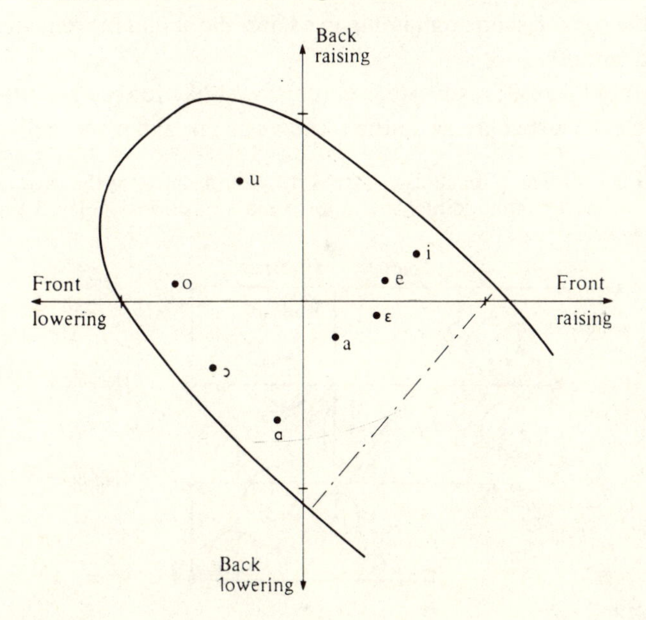

plotted as points in a plane, as shown in Fig. 2.5, whose axes represent actual tongue position. This graph is based on physical measurements. The phoneticians' version of this diagram is the 'vowel trapezium', shown in Fig. 2.6. The axes of this figure are back/front and open/close – referring to the position of the narrowest part of the oral tract. The vowel trapezium is based on a subjective assessment, and is a useful representation of relative vowel quality.

There are two additional factors which complicate the characterisation of vowels. These are lip position and degree of nasality. These constitute, in effect, two extra dimensions. Fortunately, these two variables do not seem to be used as continuous variables to any great extent. Thus vowels tend to be perceived as either nasalised or not. Similarly, only three lip positions seem to have perceptual significance; pushed-out (rounded), normal (lax), and drawn-back (tense). Thus, for each point in the graphs of Figs. 2.5 and 2.6 there are six possible conditions of the lip/nasality variable. For example, English uses the set of vowels given by the points in Fig. 2.6, without nasality, but with lip position varying from tense /i/, to rounded /u/. In contrast, French vowels not only occupy different points on the vowel diagram, but may or may not be nasalised.

Closely related to the static vowels are the diphthongs and triphthongs, which may be thought of as dynamic vowels. These consist of articulatory movements corresponding to slow changes between two (or three) vowels. They can be represented on the vowel diagram as trajectories between static vowel positions. In language they are used in exactly the same way as static vowels, the tongue shape remaining fixed and the actual movements being slow and smooth.

More rapid transitions between static, vowel-like tongue positions give another class of articulatory gestures known as the 'glides', or semi-vowels.

Fig. 2.6. The phonetician's vowel trapezium, showing the cardinal vowels (extreme points), and approximate positions of eleven English vowels.

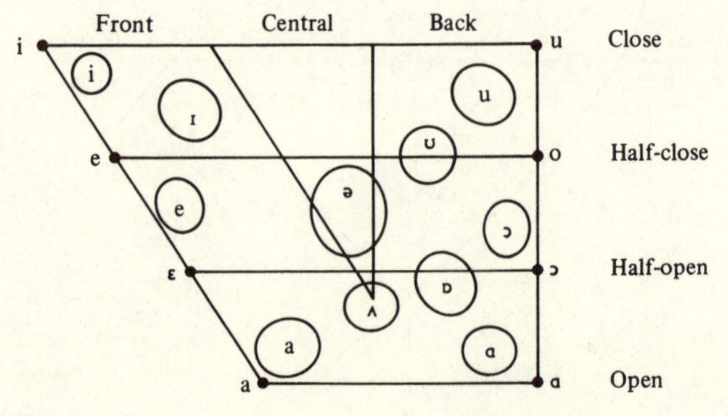

In English there are only two, /w/, and /j/. The /w/ consists of a rapid transition from a /u/ position to a /ə/ position, whereas /j/ consists of a rapid transition from /i/ to /ə/. Though one can represent these transitions on the vowel diagram as trajectories, it is really not profitable to do so. The glides are not used as vowels (probably because of the rapidity of the transition) and are more accurately thought of as transient consonants.

The final class of vowel-like articulations consists (in English) of /r/ and /l/, which are known as 'liquids'. These two gestures are essentially static but use non-vowel tongue shapes to effect partial closure of the oral tract. For /r/, the tongue tip is curled back to form a narrowing of the tract. In /l/, the tongue tip actually makes contact with the roof of the mouth, allowing air to pass by on either side. Both these gestures vary considerably in detail, depending on language, dialect and context. Again, these two articulatory gestures are more properly classified as consonants, and in language are used as such. The reason in this case seems to be that the tongue shape is in no way vowel-like.

The articulatory gestures described so far are normally uttered using voiced excitation, and for this reason these sounds are sometimes known as vocalics. It is possible, however, for all of them to be made using turbulent excitation. This is done by holding the vocal cords in a loosely constricted position, thereby causing turbulence at the glottis. The result is the whispered versions of these sounds. But in some languages and dialects whispered or unvoiced versions of /w/, /j/, /r/, and /l/ are valid articulatory gestures in their own right.

In the class of vowels, semi-vowels, and liquids the vocal tract is narrowed at a particular point, but without actually forming a turbulent constriction. When the tract is narrowed enough to cause turbulence, the resulting class of sounds is known as fricatives. In this case a static articulatory position is held, which causes turbulent air flow at the point of constriction. In English, only four fricative positions are distinguished, all toward the front of the mouth. The sound /*θ*/ is made with a constriction formed by opposing the tongue tip to the upper teeth, whereas /f/ is made by opposing the lower lip to the upper teeth. For /s/, the tongue tip is opposed to the front of the tooth-ridge, and /ʃ/ uses a grooved tongue tip to funnel air across the top of the tooth-ridge, usually accompanied by rounding of the lips.

An anomaly in English is the phoneme /h/, which usually appears before vowels. The articulatory position for this sound is that of the vowel it precedes, made with turbulent excitation at the glottis. Since /h/ usually appears together with a vowel it might be better classified as a modification of that particular vowel. This was the case in ancient Greek, where the /h/

sound was indicated by an accent above the vowel. But it is now conventional to treat it as a phoneme in its own right, and it is usually classified along with the fricatives.

The four true fricatives can all be made using voiced excitation at the same time as the turbulent constriction is formed. This gives the class of voiced fricatives, with the equivalences

/θ/—/ð/
/f/—/v/
/s/—/z/
/ʃ/—/ʒ/

The anomalous position of /h/ as a fricative now becomes clear, since it is obvious that it is impossible for it to have a voiced equivalent.

The third class of articulatory gestures to be described is the stop-plosives, so called because they involve complete closure (or stopping) of the vocal tract, and sometimes release (explosion) of air from the tract. These sounds are characterised by the fact that they terminate (stop), and/or initiate (plosive), other speech sounds. They are, therefore, highly transient in nature, and are amongst the most difficult to identify in machine recognition of speech. Though there is a continuum of points at which the vocal tract can be closed off, in any given language only a limited set of closure points are actually used. In English there are three principal points of closure, they are lips, alveolar ridge, and velum. If closure is made with the lips, then the phoneme is /p/, or /b/, depending on whether or not voicing accompanies the closure. The phoneme /b/ is thus the voiced equivalent of /p/. Similarly, /d/ is the voiced version of /t/, the closure being made at the tooth-ridge; and /g/ is the voiced version of /k/, the closure being made at the velum.

It is not at all obvious how voicing can occur when the vocal tract is closed off. What happens in these voiced stop-plosives is that the vocal cords continue to vibrate for a very brief period after the closure, the resultant flow of air into the vocal cavity being accommodated by expansion of the tract walls under the increasing pressure. This can only take place for a short time; eventually the increased pressure in the vocal tract causes glottal air flow to stop, and the vocal cords to cease vibration.

For the unvoiced stop-plosives, vocal cord vibration ceases before the closure is effected, and does not begin until after the closure is released, an event which is accompanied by a burst of 'explosive' noise. There is thus a short period of silence preceding the stop phase of the unvoiced stop-plosive, followed by a noise burst at the plosive phase. Because of their transient nature, there are many variations in the acoustic forms of these phonemes. For example, if a voiced stop-plosive occurs between two vowels, it may well be that the vocal cords will not entirely cease vibrating. Again, if a stop is followed by silence, it may or may not be 'released', that is, the stop phase may not be succeeded by the usual plosive phase. An anomalous type of stop occurs at the glottis, when the vocal cords are suddenly opened to release air in a single puff, without cord vibration. This so-called 'glottal stop' is not a standard phoneme of English, though it occurs in many dialects. However, it cannot properly be classified with the stop-plosives because it is impossible for it to have a voiced equivalent.

A class of phonemes which combine elements of stops and fricatives, is the affricates. There are two such phonemes in English, the consonants in the words 'church' and 'judge'. It is not universally agreed that these gestures are phonemes in their own right, and in the phonetic alphabet they are represented by the symbols /tʃ/ and /dʒ/. Thus, symbolically at least, they can be considered as a concatenation of an alveolar stop with an alveolar fricative, the /tʃ/ being unvoiced, and /dʒ/ being voiced. This is true, in the sense that the affricate /tʃ/, for example, combines stop and fricative articulatory gestures in the same sequence, /t/ followed by /ʃ/. But the timing of the /tʃ/ gesture is much more rapid than a simple concatenation of the stop and fricative. In actual speech there is usually no difficulty in distinguishing between the two cases – for example, between 'my chair' and 'might share'.

The final class of phonemes in English is the nasal stops. These use articulatory positions essentially the same as the stop-plosives, but with the velum lowered, so that air escapes via the nostrils. The nasal phoneme which uses lip closure is /m/, that which uses closure at the alveolar ridge is /n/, and the velar nasal stop is /ŋ/. Though the places of closure are similar, the nasal stops and the stop-plosives are very different in character. Nasal stops are not transient and can be prolonged indefinitely, neither do they have unvoiced equivalents (at least not in English).

2.4 Co-articulation

The articulatory gestures described above are those of English. They have been characterised by either static or dynamic configurations of the vocal tract, together with their usual manner of tract excitation. In order to make any description possible, they have been somewhat idealised. In natural speech, such articulations are made quickly and consecutively, so that each is modified to a certain extent by the presence of its neighbours in the sequence. This modification, which can be quite considerable, is known as 'co-articulation'. Its effect is in many ways analogous to the difference between hand printed letters and normal handwriting, in which each letter is connected to its neighbours. Depending on the speed of writing, each letter will be more or less properly formed. Similarly, in a speech utterance the degree of co-articulation will depend on the rapidity of the speech, which in turn will depend on the mode of speaking, conversational, declamatory etc. If the speech is very slow and deliberate, then co-articulatory effects will be minimal. However, in normal speech, co-articulatory transitions will make up most of the utterance, and in rapid speech some of the intended gestures will never be achieved before the articulators pass on to the next gesture. Thus the idealised articulations described above must be regarded as 'target' positions, which may or may not be reached, depending on the context and the rapidity of the speech.

There is one important difference between co-articulation in speech, and the running together of letters in handwriting. Whereas individual handwriting can often be idiosyncratic and difficult to decipher, in speech the 'smearing' of one phoneme into its neighbours is not only normal, but mandatory if the speech is not to sound stilted and unnatural. In fact, it may be argued that this overlapping of phonemes due to co-articulation is an advantage, in that it makes the speech more impervious to noise by dispersing the information in each phoneme over a greater time interval, analogous to the holographic effect. Native speakers of a language experience little difficulty in coping with the high degree of co-articulation inherent in rapid conversational speech. Since we monitor our own speech continuously, we can approach the intelligibility limit of co-articulation rather closely. This tendency makes it very difficult to synthesise speech which sounds 'natural', and even more difficult to devise speech recognition algorithms which can cope with continuous speech.

It would be wrong to give the impression that speech consists only of a sequence of articulatory gestures. The articulation of speech is accompanied by pauses, variations in loudness, and raising and lowering of voice pitch, all of which are meaningful. These variables constitute what is called 'stress' in speech, and over a longer period contribute to the effect known as 'prosody'. Indeed, without stress individual words sound rather flat, and without the correct prosodic pattern whole phrases may be unintelligible. This prosodic channel acts primarily to direct attention to the important parts of the stream of articulatory gestures, and to break them up into semantically significant units. It is, in effect, a synchronisation channel, using parameters not already employed in the articulatory channel. The three components of stress and prosody are pitch, duration, and intensity. Pitch is the fundamental frequency of vibration of the vocal cords; given that a particular articulation is voiced, the actual frequency of the voicing is a 'free' parameter available to the prosodic channel. The duration of each articulatory gesture, within limits, is variable, and taken together with intensity constitutes a stress parameter which can be used to give emphasis to certain gestures. The interaction of the prosodic and the articulatory channels usually manifests itself via changes in co-articulation. For example, a 'de-stressed' vowel will be reduced in duration, so that the tongue may not reach its target position, and thus the vowel quality may be changed, usually in the direction of the central vowel /ə/. On the other hand, if a vowel is stressed, it will affect not only the phoneme which follows it, but also the phoneme which precedes it, as the articulators anticipate the target position.

The prosodic patterns of speech also convey other, less formal information. Emotional state and attitude are readily communicated via this channel, by means not clearly understood. The process is partly unconscious and it is almost impossible not to betray strong emotions and attitudes felt whilst speaking. Newsreaders have to be trained to convey a rather neutral attitude to the material they are required to read. Actors, on the other hand, develop the skill to communicate quite complex emotional states via this channel.

Assigning prosodic patterns to synthetic speech has been the subject of considerable research. Simple rules exist for assigning stress and intonation to achieve the synchronisation function discussed above. This can be shown to have a dramatic effect on the naturalness of the speech, since without it the sequences of articulations are not only flat and monotonous, but usually unintelligible.

2.5 Spectrograms

So far human speech has been characterised as a set of more or less formalised articulatory movements of the vocal tract. It is also possible to describe speech sounds in terms of their acoustic features, and this is the task of 'acoustic phonetics'. The problem in characterising the units of speech in terms of acoustic parameters is that these descriptions will vary considerably from speaker to speaker, and from utterance to utterance. It is, in fact, impossible for the same speaker to say the same thing twice in exactly the same way. Thus acoustic phonetics proceeds by giving general descriptions of the sounds, and a set of rules which indicate how such sounds may vary in different contexts and with different speakers.

It has long been known that the human ear detects frequency spectra of sounds, rather than their actual time waveforms. The basilar membrane of the inner ear performs an approximate short-time frequency-transform of sound pressure, and it is this information which is presented to the brain via the auditory nerve. In consequence, the speech spectrograph was

Fig. 2.7. (*a*) Spectrogram of 'seat', narrow band. (*b*) Spectrogram of 'seat' wide band.

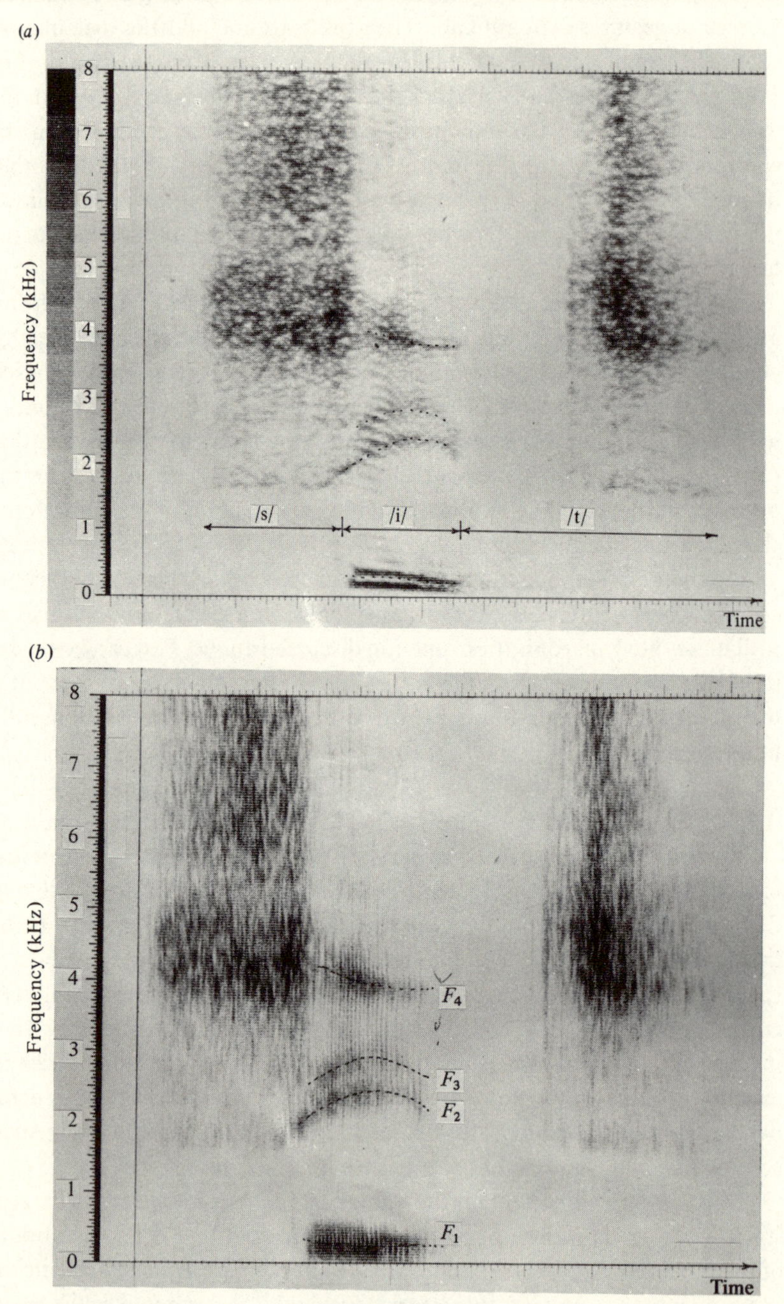

invented to convert the time waveform of sounds to short-time spectra, and to display this three-dimensional information (time, frequency and intensity) as an energy-density map. The spectrogram has thus become the accepted method of describing the acoustic features of speech. This time/frequency/intensity display shows the spectral distribution of energy throughout an utterance in a dramatic and meaningful way. The result is a sound 'picture', meaningful in the sense that similar sounds give similar spectrograms. That is, utterances which sound the same to the human ear look the same, on the spectrogram, to the human eye. In fact, neither sounds nor spectrograms are ever exactly the same, and the degree of similarity between two utterances is extremely difficult to quantify. Acoustic phonetics avoids this problem by describing speech sounds in terms of spectrograms, and leaving the problem of similarity to be resolved by the human eye and brain.

Fig. 2.7 shows spectrograms of the word 'seat', phonetically /sit/. The vertical axis represents frequency, covering the range 0–8 kHz, and the horizontal axis is proportional to time. The 'blackness' of the spectrogram at each point is proportional to the logarithm of the energy of the speech at that particular frequency and at that particular time. The intensity range of the 'blackness' (or grey scale) covers about 40 dB. In the spectrogram of Fig. 2.7(*a*), the analysis filter has been set to narrow band (50 Hz), so that the frequency resolution is high. The time resolution (the reciprocal of bandwidth), however, is low. The result is that the pitch harmonics of the utterance are resolved quite clearly on the frequency axis, and appear as horizontal bands of energy. In contrast, in Fig. 2.7(*b*), the analysis filter is wideband (200 Hz), thus the frequency resolution is low, and the time resolution is high. In this case the individual pitch harmonics are not distinguishable on the frequency axis, but the impulsive nature of the time waveform is resolved on the time axis, and is manifest as vertical bands of energy.

Cross-sections of these spectrograms at particular time instances are illustrated in Fig. 2.8. The harmonic structure of the vowel spectrum is clearly visible in Fig. 2.8(*a*), which shows a cross-section from the narrow band spectrogram of Fig. 2.7(*a*). The high frequency concentration of energy in the fricative /s/ is shown in the cross-section of Fig. 2.8(*b*), which is taken from the wideband spectrogram of Fig. 2.7(*b*).

The harmonic structure of the voiced spectrum, shown in Fig. 2.8(*a*), is shaped by an 'envelope' function, characteristic of the resonances of the vocal tract. The peaks in this spectral envelope are known as 'formants'. In vowels, there are usually four to five formants in the frequency band of interest (up to 5 kHz), though for a given speaker three formants are usually enough to characterise a particular vowel. Other speech sounds may also be characterised by formants, but the situation is usually more complicated. The fricative /s/ of 'seat' appears on the left of the spectrograms of Fig. 2.7 as wideband energy. There are no horizontal or vertical striations because the sound is not voiced. Its spectral cross-section, shown in Fig. 2.8(*b*), is characteristic of wideband noise shaped by high frequency formants. However, the detailed structure of the formants is not as perceptually significant as it is for voiced sounds.

The /t/ in 'seat' appears on the right of the spectrogram as a silent gap, which is the closure of the stop, followed by a burst of wideband energy, which is the turbulent noise emitted when the stop is released. The formants characteristic of the vowel /i/ are to be seen in the centre of the voiced region. As the vocal tract changes shape in the transition from the fricative /s/ to the vowel /i/, and from /i/ to the stop /t/, the formant frequencies also change position. Perceptual clues to the identity of the fricative /s/ and the

Fig. 2.8. (*a*) Cross-section of spectrogram of Fig 2.7(*a*), from vowel portion of 'seat'. (*b*) Cross-section of spectrogram of Fig 2.7(*b*), from fricative in 'seat'.

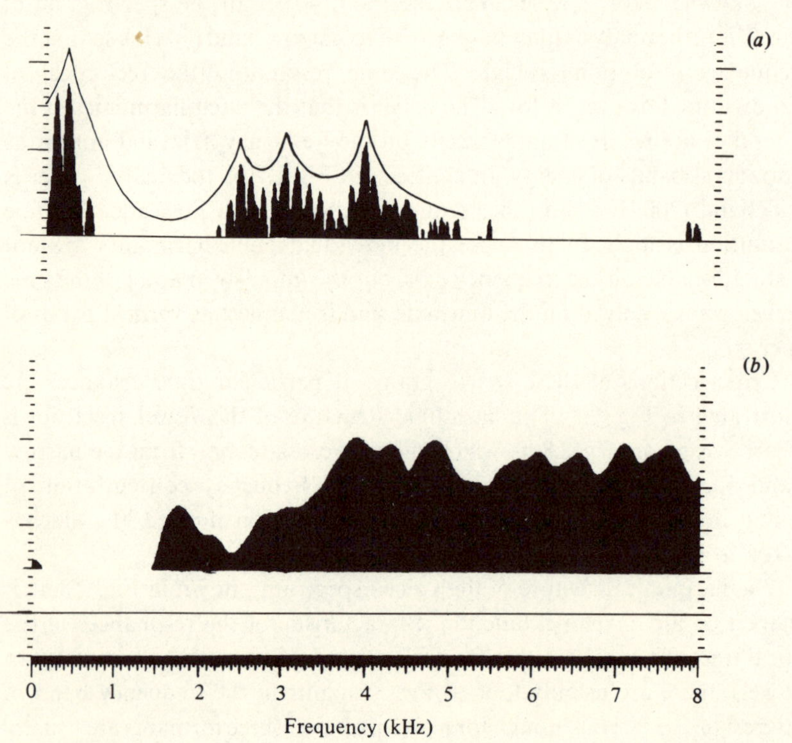

stop /t/ are present in these formant transitions, which are the acoustic correlates of co-articulation.

Spectrograms are indispensable aids in analysing both human and synthetic speech. Usually, when a spectrogram of synthetic speech resembles that of real speech, the synthetic and the real speech will sound the same. The acoustic characteristics of the English speech sounds which follow will, therefore, be given in terms of descriptions of spectrograms. Since most of the information in the speech signal is contained in the lower part of the spectrum, the following spectrograms will only cover the range 0–4 kHz. Also, because time resolution is generally considered to be more important than frequency resolution, the spectograms will all be wideband.

The classification of speech at the acoustic level will be slightly different from that at the articulatory level. Thus, the classes of articulations, vowels, diphthongs, semi-vowels and liquids, can be grouped together under the term 'approximants', since they are all approximate closures of the vocal tract and use voiced excitation. The spectrograms of approximants are characterised by voicing (without turbulence), and the presence of formants. Fig. 2.9 shows the spectrogram of the nonsense utterance /jər waul/, which contains one vowel, one diphthong, both liquids and both semi-vowels. The central vowel /ə/ has formants at frequencies $F_1 = 450$ Hz, $F_2 = 1400$ Hz, $F_3 = 2400$ Hz and $F_4 = 3400$ Hz. For a given speaker three formants are usually enough to characterise the vowel. Typical formant frequencies for the vowels of a male speaker are given in the table of Fig. 2.10.

The diphthong /au/ changes quality from /a/ ($F_1 = 700$ Hz, $F_2 = 1200$ Hz, $F_3 = 2700$ Hz) to /u/ ($F_1 = 500$ Hz, $F_2 = 1000$ Hz, $F_3 = 2500$ Hz) over a period of several hundred milliseconds. This change is smooth and diphthongs can be described in terms of the formants of their constituent vowels. By contrast, the semi-vowel /j/ begins with formants characteristic of /i/, but changes quickly to a central position. The change is

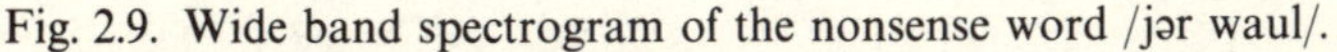
Fig. 2.9. Wide band spectrogram of the nonsense word /jər waul/.

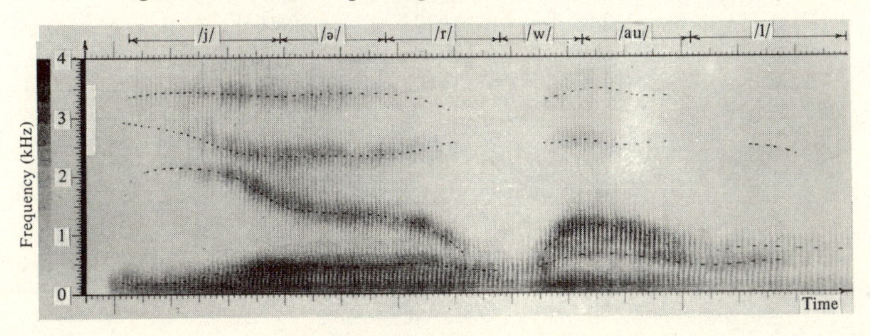

quite rapid and there is no steady state. Similarly, /w/ changes from /u/ to /ə/ very quickly. Both /j/ and /w/ can be described in terms of the formants which characterise the beginning and end of their trajectories, and also by the speed of the transition.

The liquids /r/ and /l/ are nominally static articulations, and as such may be characterised by formant frequencies. However, they are usually made quite quickly and they appear in the spectrogram of Fig. 2.9 at the ends of vowels, /r/ after /ə/, and /l/ after /u/. Liquids generally have less energy than vowels and exist for shorter duration, so that they may appear as 'inflections' at the beginning or end of vowel formant trajectories.

The fricatives, that is, sounds made with turbulent excitation, appear on the spectrogram as broad smudges. The random noise characteristic of turbulent air flow has a wide frequency spectrum, which is shaped by the resonances of the vocal tract. The resonant cavities involved are those forward of the turbulent constriction, so that the resonant frequencies are higher then vowel formants. Fig. 2.11 shows spectrograms of the nonsense syllables, /θəʒ/, /fəz/, /səv/ and /ʃəð/. The pure fricatives beginning each syllable are clearly visible on the spectrum, though /θ/ and /f/ are weaker than /s/ and /ʃ/. Several factors permit them to be distinguished from each other: one is the distribution of energy over the spectrum, another is the total amount of energy. An additional factor is the formant transitions from the fricative to the following vowel; this is especially obvious in Fig. 2.11(*d*), for the phoneme /ʃ/.

The voiced fricatives /ð/, /v/, /z/ and /ʒ/ appear at the end of the syllables in Fig. 2.11. They are essentially the same as their unvoiced counterparts, except that they have voicing superimposed in the low frequency band. This

Fig. 2.10. Table of formant frequencies of English vowels for typical male speaker in Hertz.

Vowel	F_1	F_2	F_3
i	280	2620	3380
I	360	2220	2960
e	600	2060	2840
æ	800	1760	2500
ə	560	1480	2520
ɑ	740	1180	2640
ʌ	760	1370	2500
ɒ	560	920	2560
ɔ	480	740	2620
ʊ	380	940	2300
u	320	920	2200

Fig. 2.11. Wide band spectrograms of nonsense syllables; (*a*) /θəʒ/, (*b*) /fəz/, (*c*) /səv/, and (*d*) /ʃəð/.

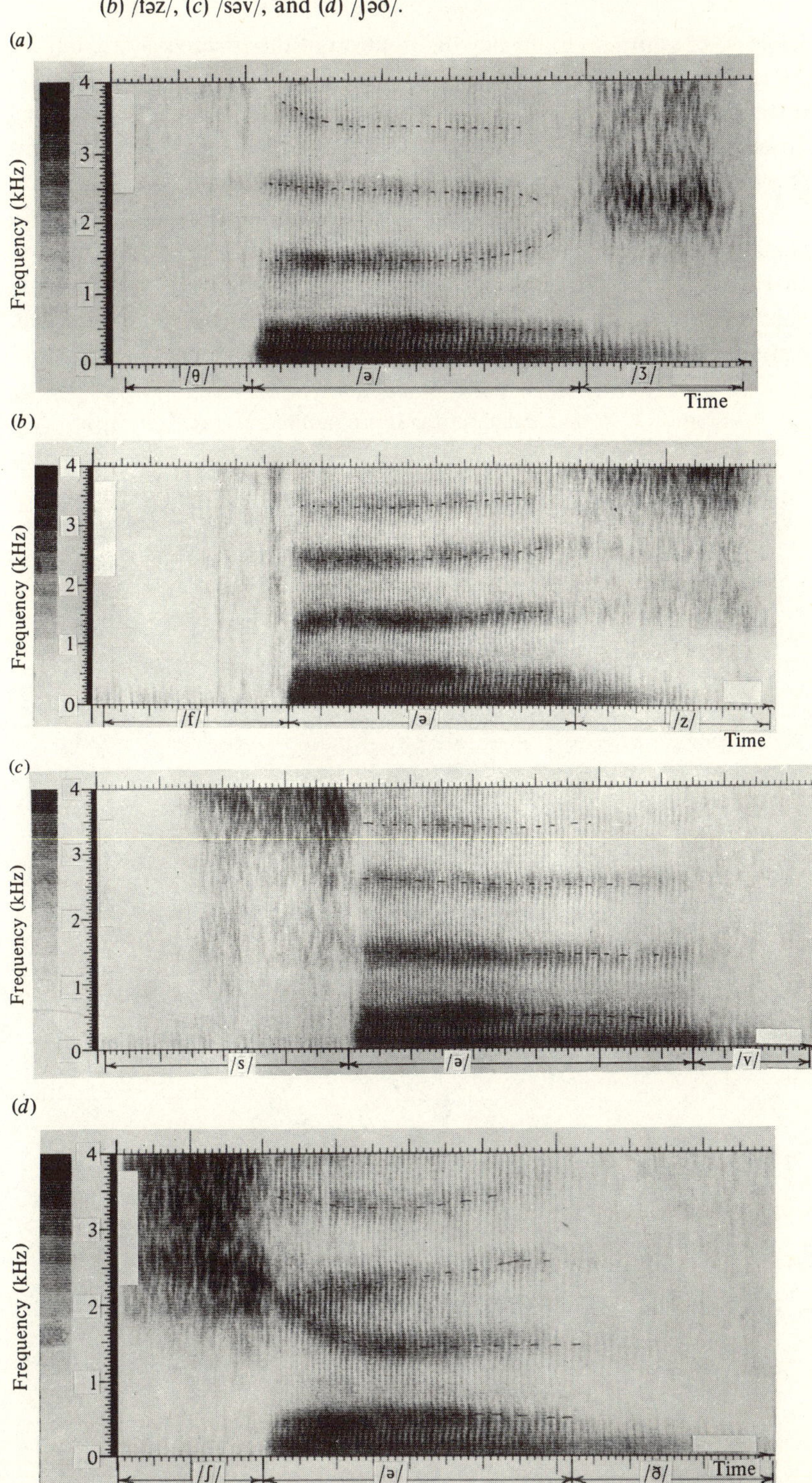

tends to obscure the frequency distribution of the fricative noise, but for these sounds the ratio of low to high frequency energy is a good distinguishing feature. Again, the formant transitions from the preceding vowel provide significant clues to the identity of the voiced fricative. This can be seen most clearly in Fig. 2.11 (*a*), for the phoneme /ʒ/.

Finally, Fig. 2.12 shows spectrograms of the nonsense words /bapam/, /datan/ and /gakaŋ/, which contain the voiced, unvoiced and nasal stops. In each case the unvoiced stop-plosive appears between two strongly voiced vowels, and their main characteristic is the silent gap, which distinguishes them from their voiced counterparts. The voiced stop-

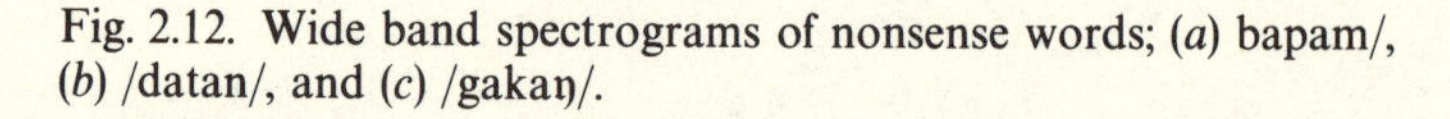

Fig. 2.12. Wide band spectrograms of nonsense words; (*a*) bapam/, (*b*) /datan/, and (*c*) /gakaŋ/.

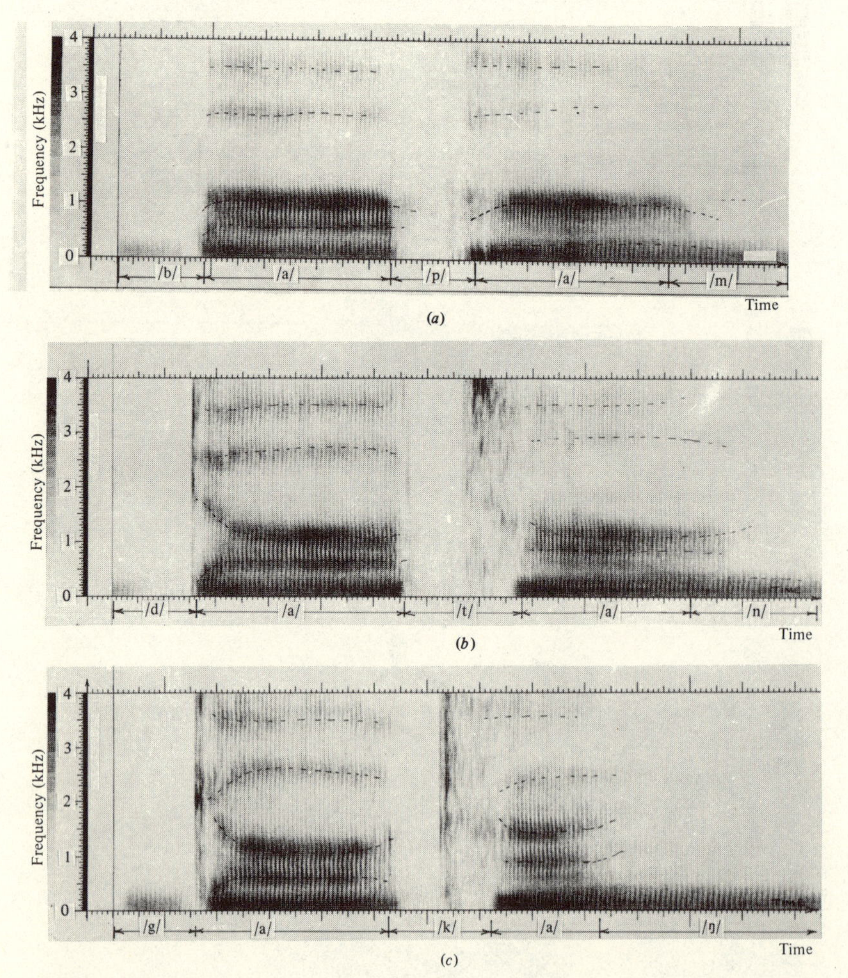

plosives appear at the beginning of the nonsense words, where closure is accompanied by low-amplitude, low-frequency, voiced energy. The plosive noise burst on release is similar in the voiced and unvoiced cases. The nasal stops /m/, /n/ and /ŋ/, appearing at the ends of the nonsense words, are easily distinguishable from the stop-plosives by the presence of a high-energy low-frequency formants, and absence of high-frequency noise.

For all stops the place of closure is mainly characterised by the formant transitions into and out of the stop. This can be seen most clearly for the unvoiced stops, sandwiched between two vowels in Fig. 2.12. If the formant trajectories of the vowels are projected into the silent gap, then their locus in each case will be an indication of the point of closure. Another important clue which helps to distinguish voiced from unvoiced stop-plosives, is voice-onset time. This is the duration of the pause between the release of the stop and the beginning of voicing. Voice-onset times can vary from zero for voiced stops to several tens of milliseconds for aspirated stops. Thus, the characteristic acoustic features which distinguish the stops from each other are to be found in the transitions to and from adjacent sounds. This has important consequences for synthetic speech.

The descriptions of the sounds given above are much simplified, and only their main features have been outlined. In natural speech, the sounds are affected to a much greater extent by co-articulation, and the presence of a particular phoneme is often difficult to detect from a spectrogram. There are essentially two problems in deciphering spectrograms – detecting the presence of a phonetic unit (segmenting), and identifying such units (labelling). Several research workers have shown that, with training, it is possible to 'read' spectrograms to a fair degree of accuracy. However, this 'reading' is only possible because of the pattern recognition ability of the human eye and brain. Translating spectrograms automatically by computer program is still a subject of considerable research.

The consequence of the almost total context-dependence of speech sounds is that speech synthesis is not as simple as might first be thought. It is not possible simply to concatenate sequences of basic sound units to produce an utterance. The effects of co-articulation must be taken into account and each phoneme must be carefully blended with its neighbours in the sequence. Furthermore, the effects of prosody will overlay and modify the effects of co-articulation. Thus, an acoustic description of speech as a set of basic sound units is in no way satisfactory, without sets of rules which describe how these units are modified, by prosody and by co-articulation. The synthesis of speech from phonetic text relies on such rules, and the main advance of speech science in recent years has been in the discovery of adequate phonological and prosodic rules.

3 *Articulatory synthesis*

Modelling the speech production mechanism

'It is clear that in speaking the lungs attract air and expel it; it is also clear that the air is agitated by the glottis, as if by a reed, and that it makes a sound. Finally, it is clear that the mouth and tongue change position with each note, thus forming different obstacles to the expulsion of the air which is producing the sound. In other words, they offer it openings which are sometimes larger, sometimes smaller, and of various shapes. In this way I determined that speech or articulation is nothing other than voicing which passes through different apertures.'
Von Kempelen, 1791

3.1 Introduction

Articulatory synthesis is the production of speech sounds using a model of the vocal tract which, directly or indirectly, simulates the movements of the speech articulators. Models of this kind may have varying degrees of complexity, and can be either analogue or digital circuits or, more usually, computational models in the form of computer programs.

At the present time the complexity of articulatory synthesis is such that it is not yet appropriate for commercial application. Articulatory synthesisers are, however, ideal for research into articulatory phonetics and speech research in general; and, for the engineer interested in low bit rate communication, they hold out the ultimate promise of high quality, natural-sounding speech from simple phonetic text input.

The strategy of articulatory synthesis is that by modelling the vocal tract directly, the model will require only simple control signals to give good quality speech, the assumption being that co-articulation effects, so difficult to formalise as rules, will occur naturally – essentially as they do in real speech.

The parameters which are used to represent the shape of the vocal tract are cross-sectional areas at discrete points along its length. Unfortunately, to represent adequately all the variation in the vocal tract shape, the areas need to be specified at 20 to 40 separate points. This implies that there are 20 to 40 degrees of freedom in the system. Fortunately, this is not true and the large number of cross-sectional areas can be derived from as few as seven independent variables, which span the whole range of realistic vocal tract shapes. As might be expected, these independent variables turn out to be approximately equivalent to the movements of the speech articulators.

To simulate the movement of the vocal tract, the area function must change with time. Since the shape of the tract changes relatively slowly, the area function need only be updated every 20–50 ms. However, the trajectory of these changes in time is very important, and imperfections in the articulatory dynamics can have quite severe effects on speech quality.

As well as a vocal tract area function, articulatory synthesis requires an exact specification of the vocal tract excitation. This is complicated in the vocal tract model for two reasons. Voiced excitation ought to be realistic in that the vocal cords should open and close, and unvoiced excitation must be included at the point of constriction. To be in any way complete, an excitation model should take into account the air flow conditions in the vocal tract.

A properly constructed articulatory model is capable of reproducing all the perceptually relevant effects that occur in real speech. The only obstacle therefore to synthesising speech indistinguishable from human speech is in the specification of appropriate signals to control the tract shape and the excitation. Methods of specifying area functions are now at an advanced stage; they promise not only high quality speech from low bit rate control parameters, but also considerable insight into articulatory dynamics.

3.2 Basic acoustical analysis

The vocal tract is an acoustic system of varying shape and cross-sectional area, with a right-angled bend at the junction of the pharynx and the oral cavities. There is a nasal side branch beginning at the top of the pharynx, of fixed dimensions but variable coupling. This system can be represented for the purposes of acoustic analysis, and without too much inaccuracy, by the two-tube model shown in Fig. 3.1. Here the oral cavity is

Fig. 3.1. Two-tube model of the vocal tract.

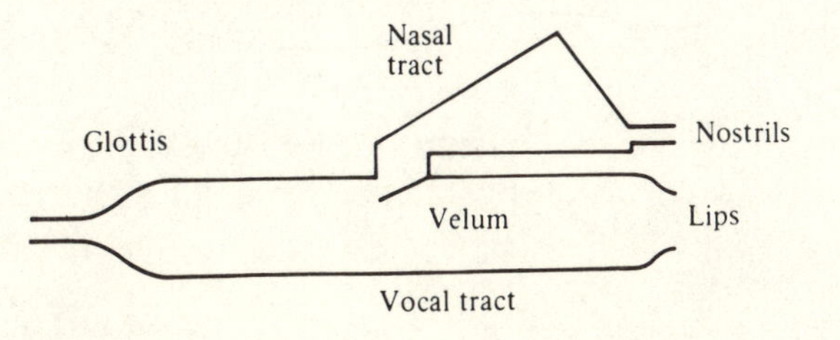

represented as a straight tube of varying cross-sectional area but of fixed shape (usually circular). The nasal cavity is represented as a tube of fixed shape and fixed cross-sectional area, variably coupled to the oral tract. Since the nasal tract is of fixed dimensions it can be represented by its input/output characteristic, which, once calculated, can be used in the model without further modification.

The oral tract shape, however, varies in time due to the movements of the articulators, and its acoustic characteristics must be calculated or simulated at a rate in sympathy with these changes. Assuming that the tube is circular in cross-section, the essential variable is then its cross-sectional area, $A(x)$, where x is the distance along the vocal tract from the glottis. The technique is to assume that $A(x)$ is fixed for the short interval during which the analysis is to take place. Since the movements of the articulators are somewhat slower than the lowest frequency of interest in speech (about 50 Hz), $A(x)$ may be assumed constant over an analysis frame up to 50 ms in duration.

Another simplifying assumption is that only plane wave propagation along the length of the tube is significant. This means that the width of the tract must be smaller than the wavelength of the highest frequency of interest. Since the largest lateral dimension is about 3.5 cm the analysis ought to be valid below about 10 kHz. This is quite adequate for speech where 5 kHz is considered to be an appropriate bandwidth. The vocal tract is thus approximated as an acoustic transmission line, discrete in both length and time. Such a system is analogous in many ways to an electrical transmission line, and this analogy is emphasised in the following analysis.

Consider the elemental section of an acoustic tube at the point x, of length Δx, and cross-sectional area $A(x)$, as shown in Fig. 3.2. Let the pressure at this point be $P(x, t)$ and the flow, in terms of volume velocity, be $U(x, t)$. Both $P(x,t)$ and $U(x, t)$ are functions of time, t, as well as x. For the

Fig. 3.2. Acoustic model of the oral tract, showing an elemental section of length Δx.

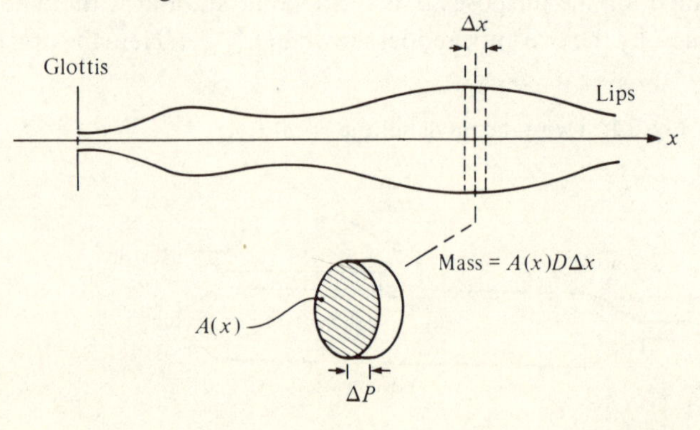

sake of clarity $A(x)$, $P(x, t)$ and $U(x, t)$ will be written simply as A, P and U. The purpose of this analysis is to discover the relationships between the parameters A, P and U, and the independent variables x and t.

Two important relationships may be developed from two basic physical properties of the air in the tube – its momentum and its compressibility. The mass of air in the elemental section will be accelerated by the force acting upon it, according to the relation

$$\text{force} = \text{mass} \times \text{acceleration}$$

The force acting across the section in the direction of x is given by $A\Delta P$, where ΔP is the pressure difference across Δx. Thus

$$A\Delta P = A\Delta x \cdot D\frac{\partial}{\partial t}\left(\frac{U}{A}\right)$$

where D is the density of the air. Since A is not being considered as a function of time, we have

$$\frac{\Delta P}{\Delta x} \rightarrow \frac{\partial P}{\partial x} = \frac{D}{A}\frac{\partial U}{\partial t}$$

The relationship which expresses the compressibility of the air in the elemental section is the gas law

$$\text{pressure}\left(\frac{\text{volume}}{\text{mass}}\right)^n = \text{constant}$$

where n is the ratio of specific heats. Now the volume of the elemental section is fixed, so that compression occurs due to the build-up of mass in the section. The rate of increase of mass is thus $\Delta U \cdot D$. It can be shown that the rate of change of pressure, $\partial P/\partial t$, due to this effect is given by

$$\frac{\Delta U}{\Delta x} \rightarrow \frac{\partial U}{\partial x} = \frac{A}{nP_0}\frac{\partial P}{\partial t}$$

where P_0 is the atmospheric pressure. This equation is valid only if the changes in P are small compared with P_0. Under this restriction the density

Fig. 3.3 Elemental section of a lossless electrical transmission line.

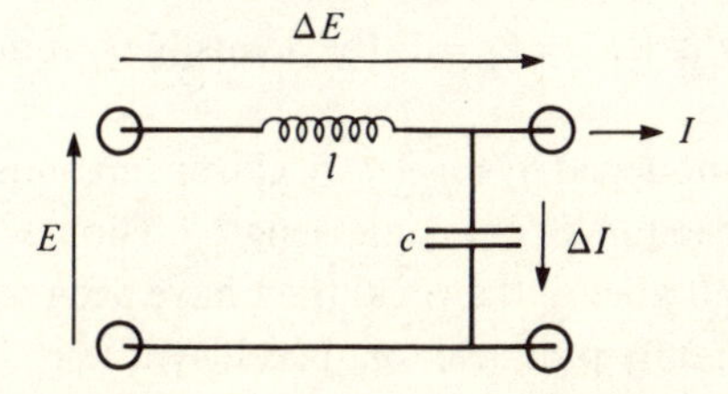

D is also substantially constant, and the equation can be rewritten

$$\frac{\partial U}{\partial x}=\frac{A}{DV^2}\frac{\partial P}{\partial t}$$

where V is the velocity of sound in air, which may also be assumed constant.

Consider now the elemental section of a lossless electrical transmission line of length Δx, as shown in Fig. 3.3. The elemental series inductance

$$l=L(x)\Delta x$$

where $L(x)$ is the inductance per unit length of the line. The elemental shunt capacitance

$$c=C(x)\Delta x$$

where $C(x)$ is the capacitance per unit length of the line. Let the voltage and current at the point x on the line be $E(x,t)$ and $I(x,t)$, respectively. Again, for the sake of clarity $L(x)$, $C(x)$, $E(x,t)$ and $I(x,t)$ will be written simply as L, C, E and I.

The voltage drop along the line ΔE is given by $\Delta E = L(\partial I/\partial t)\Delta x$, thus

$$\frac{\Delta E}{\Delta x}\rightarrow\frac{\partial E}{\partial x}=L\frac{\partial I}{\partial t}$$

and the current, ΔI, through the capacitor is given by $\Delta I = C(\partial E/\partial t)\Delta x$, thus

$$\frac{\Delta I}{\Delta x}\rightarrow\frac{\partial I}{\partial x}=C\frac{\partial E}{\partial t}$$

The analogy with the acoustic phenomena is obvious. It is thus possible to speak of an 'acoustic inductance' L_a, equivalent to D/A, and an 'acoustic capacitance' C_a, equivalent to A/DV^2. Introducing constants k_1 and k_2, to scale pressure to voltage, and volume velocity to current, these equivalences can be made direct,

$$L_a=\frac{k_1 D}{k_2 A}=\frac{D}{kA}$$

and

$$C_a=\frac{k_2 A}{k_1 D V^2}=\frac{kA}{DV^2}$$

where $k=k_2/k_1$. In the case where $k_1=k_2=1$, the acoustic quantities are said to be normalised.

The vocal tract can thus be modelled by an electrical transmission line of varying inductance and capacitance per unit length. Though many electrical transmission line analogues of the vocal tract have been reported in the literature, they are not really practical for speech synthesis, for two main reasons. The first is that it is very difficult to vary real inductors and capacitors quickly enough to simulate articulatory movements. The second reason is that acoustic losses need to be taken into account, and it is difficult to simulate them dynamically, in a realistic manner. However, static transmission line analogues have been built which give excellent simulations of static vocal tract configurations – essentially vowels. In these models the inductances and capacitances are varied in switched steps, and the acoustic losses are simulated by series and/or shunt resistances which are adjusted by hand to realistic values.

Acoustic losses can be accounted for in a transmission line analogue by including resistors in the circuit. However, there is no simple relationship between the resistance values and the vocal tract area function, especially at low values of $A(x)$. At constrictions in the tract, there is the additional problem that turbulent flow may take place, giving a dramatic increase in losses and also random noise excitation. In this case there is no natural electrical or electronic analogue, and these effects are very difficult to simulate. In fact, simulating losses is a major problem in articulatory modelling, and it is therefore worthwhile discussing them in more detail.

3.3 Acoustic losses

Extending the transmission line analogy, it is possible to represent acoustic losses due to laminar flow by series and shunt resistances in the basic transmission line section, as shown in Fig. 3.4. The series resistance represents losses due to viscous drag in which the energy loss is proportional to the square of volume velocity, and the shunt conductance represents loss due to heat conduction which is proportional to pressure squared.

The viscous loss mechanism is one of friction. If the layer of air contiguous with the tube wall is assumed to be stationary, and the air in the centre of the tube to be moving with velocity U, then there will be a radial gradient of velocity across the section of the tube. This is illustrated in Fig. 3.5. The viscous friction may be thought of as taking place between

Fig. 3.4. Elemental section of a lossy electrical transmission line.

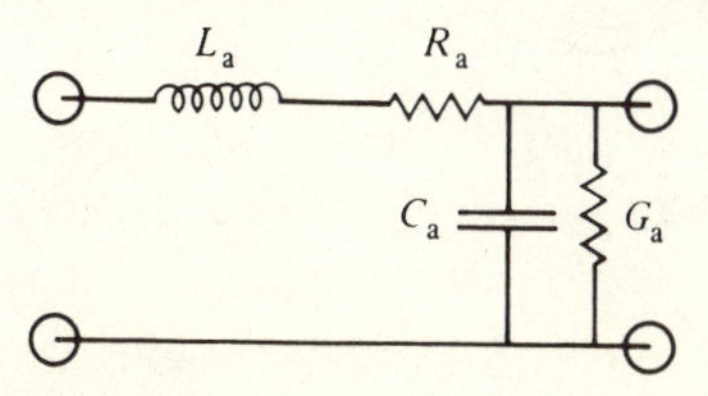

concentric laminar rings of air, each moving with a slightly different velocity from that of its neighbours. The average power loss due to this effect is dependent on m, the viscosity of air, and has been shown by Flanagan (1972) to be

$$\frac{\bar{U}^2 S}{2A^2}\left(\frac{\omega D m}{2}\right)^{1/2}\Delta x$$

where S is the circumference of the tube and ω is the radian frequency variable. This analysis is in the frequency domain and assumes that volume velocity is sinusoidal, that is, $U(t) = \bar{U}\sin\omega t$. The analogous situation in an electrical transmission line with current, $I(t) = \bar{I}\sin\omega t$, and series resistance per unit length, R, is an average power loss of $\frac{1}{2}\bar{I}^2 R\Delta x$. The normalised 'acoustic resistance' per unit length is thus

$$R_\mathrm{a} = \frac{S}{A^2}\left(\frac{\omega D m}{2}\right)^{1/2}$$

The dependence of R_a on frequency ω denotes that it is not a pure resistance in the electrical sense. Also, the dependence on S, the tube circumference, conflicts with the assumption that the cross-sectional shape of the acoustic tube is irrelevant.

The shunt conductance in a transmission line gives losses which are proportional to the square of the voltage. The equivalent loss in an acoustic tube is the loss due to heat conduction at the tube walls. This is a difficult process to visualise, since it might be assumed that the air and the tube would be at one uniform temperature. But this is only true on a macroscopic scale. The rapid and essentially adiabatic variations in

Fig. 3.5. Representation of viscous air flow in round-section tube, illustrating the radial gradient of volume velocity.

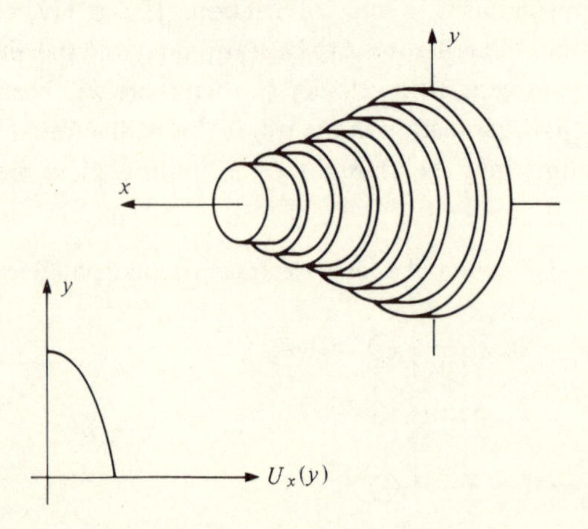

pressure cause simultaneous variations in temperature. The physical law involved in this process is the adiabatic gas law

temperature = pressure × volume × constant

If the walls of the tube were perfectly insulating (thermally), then these temperature variations would be uniform across the tube section. But since the wall will conduct heat to and from the air in the tube, there will be a temperature gradient across the section and a net heat loss. Assuming the tube wall to be perfectly conducting, that is, at constant temperature, the energy loss will depend only on the thermal conductivity, l, and the specific heat, h, of the air. Flanagan (1972) has shown that the average power loss due to this effect is given by

$$\frac{\bar{P}^2 S(n-1)}{2DV^2}\left(\frac{l\omega}{2hD}\right)^{1/2}\Delta x$$

Again this assumes a sinusoidal pressure variation $P(t) = \bar{P}\sin\omega t$. The analogous situation in an electrical transmission line with sinusoidal voltage $E(t) = \bar{E}\sin\omega t$ is that power loss is given by $\frac{1}{2}\bar{E}^2 G\Delta x$, where G is the shunt conductance per unit length of the line. The normalised 'acoustic conductance' is thus

$$G_a = \frac{S(n-1)}{DV^2}\left(\frac{l\omega}{2hD}\right)^{1/2}$$

The analogy is again unsatisfactory, since G is a function of frequency ω, and S, the circumference of the tube.

It can be seen that in the case of inductance and capacitance the analogy between an electrical transmission line and an acoustic tube is very good. It becomes less useful when extended to series and shunt losses, since in these cases the electrical elements must be frequency-dependent, and are no longer simple functions of the tube cross-sectional area. For modelling turbulent flow the analogy breaks down completely, since there is no comparable electrical phenomenon. However, Stevens, Kasowski & Fant (1953) have shown that the transmission line analogy can still be maintained, if a noise generator is inserted in the line at the point of constriction. The power of the noise source and its internal resistance will depend on the constriction area and the volume velocity. Flanagan (1972) gives approximate expressions for the constriction resistance

$$R_c = \frac{DU}{2A^2}$$

and the r.m.s. noise pressure

$$P_c = H_1 A^2 U^2 W^2 - H_2, \quad P_c > 0$$

where W is the width of the constriction (assuming a slit), and H_1, and H_2 are constants. H_2 is a threshold below which the turbulent noise is negligible. In electrical terms R_c is a current-dependent resistance, and P_c is a current-dependent voltage source. There are no electrical devices which correspond to these specifications, so that modelling is only possible by careful adjustment of variable components corresponding to R_c and P_c.

So far in this analysis, the acoustic tube has been considered to be hard-walled. In the real vocal tract this assumption is not valid. Not only will the walls vibrate in sympathy with the pressure wave, but with yielding walls the actual volume of the tube will increase with the steady pressure in the tract. This effect is quite significant for voiced stops, when the tract is suddenly cut off and voicing continues for a short time. If the walls of the model are unyielding the pressure will build up too quickly, and the vocal cords will stop vibrating almost immediately. Furthermore, during the closure, sound escapes from the tract via vibration of the tube walls. Both these effects can be simulated in a transmission line analogue, by using a 'wall impedance' Z_w, consisting of the circuit elements shown in Fig. 3.6. Here, wall compliance, mass and damping are modelled by a series inductance, capacitance and resistance circuit, in parallel with the line. That part of the wall vibration which is communicated to the outside air is represented as a voltage E_0 across part of the resistance and inductance. The total value of energy loss involved in this effect is not significant compared with the series and shunt losses, R_a and G_a. However, the effect of wall compliance and radiation during voiced stops is significant enough to warrant their inclusion in a really sophisticated vocal tract model.

3.4 Nasal tract

The nasal tract is coupled to the vocal tract at the pharynx–mouth junction, as illustrated in Fig. 3.7(*a*). The velum, when lowered, admits air to the nasal cavity, but does not fully obstruct the oral cavity. However, when raised, the velum cuts off the nasal tract completely. Since the velum is essentially a flap between the nasal and oral cavities, its position affects the cross-sectional area of both cavities in its immediate

Fig. 3.6. Electrical model of wall compliance; E_0 represents the sound emitted via wall vibration; Z_w is wall impedance.

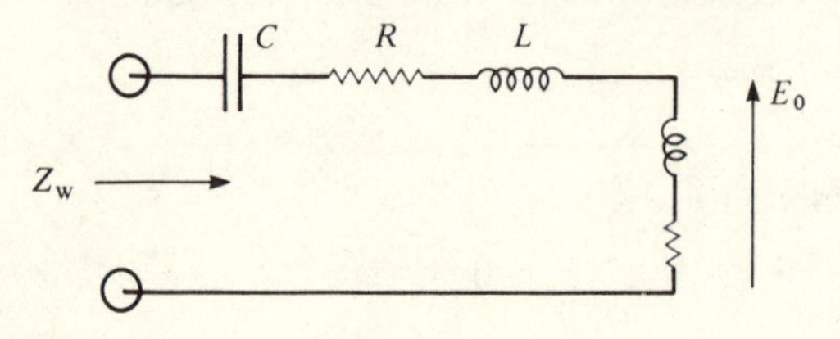

vicinity. A reasonable model in electrical terms is two parallel transmission lines coupled together as shown in Fig. 3.7(*b*). Since the nasal tract forward of the velum is of fixed dimensions (volume about 60 cm^3), it can be represented by a transmission network with fixed parameters. This can be a short transmission line or an approximation in terms of inductors, capacitors and resistors. Since the nasal tract is somewhat convoluted in shape, and has a large surface area compared with its volume, its losses will be greater than those of the oral tract. Essentially, it contributes a single, well damped, low-frequency formant to the speech spectrum.

In certain articulations, spectral zeros will also be manifest. These appear

Fig 3.7. (*a*) Section through the head showing connection of the oral and nasal cavities. (*b*) Transmission line model of the oral and nasal tract connection.

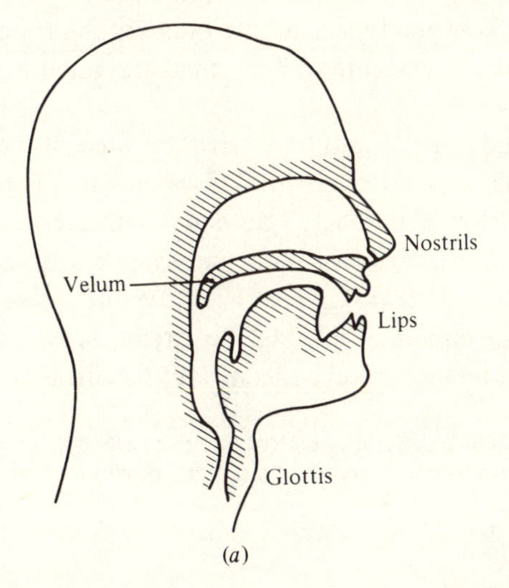

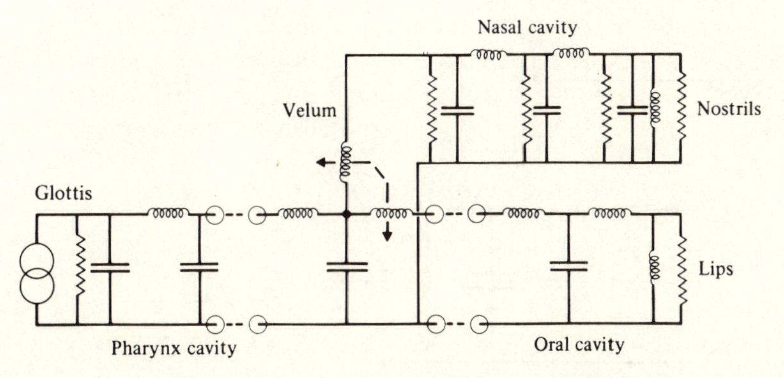

in nasal stops when the oral cavity is completely cut off and the only air flow is via the nostrils. In this situation the stopped oral cavity acts as an open circuited transmission line, and has zeros of input impedance for frequencies at which the line length corresponds to odd quarter wavelengths. This impedance appears in parallel with the nasal tract and absorbs energy at these frequencies, giving zeros in the transmission characteristic.

3.5 Acoustic radiation

Acoustic energy escapes from the vocal tract via the lips and, in the case of nasal sounds, via the nostrils. This loss of energy from the system appears as a resistive load, and in the transmission line model can be represented by resistance and other frequency-dependent components. An approximate model for the acoustic analysis is that of a circular aperture in a sphere. An even simpler model is a circular aperture in an infinite plane. Since head radius is about 9 cm and typical mouth radius is about 1 cm, the infinite plane model is not too inaccurate. For sounds radiated from the nostrils it is even better.

The infinite plane model is usually preferred because it yields a particularly simple electrical equivalent circuit. This is shown in Fig. 3.8(*a*); it is an inductor and resistor in parallel. The value of the resistance in normalised ohms is 1.441. The value of the inductance varies with radius of the aperture, a, and is given by $0.8488a/V$, where V is the velocity of sound. A more elaborate equivalent circuit which fits the circular-aperture-in-a-sphere model has been proposed by Stevens *et al.* (1953). This is shown in

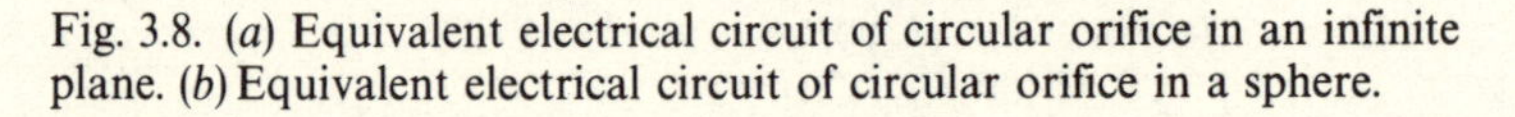
Fig. 3.8. (*a*) Equivalent electrical circuit of circular orifice in an infinite plane. (*b*) Equivalent electrical circuit of circular orifice in a sphere.

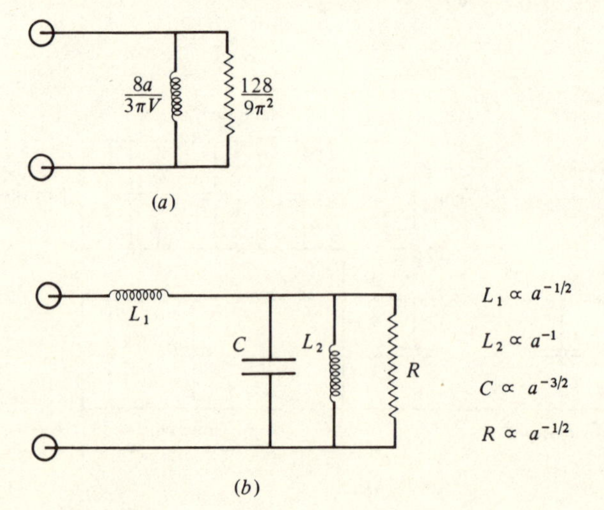

Fig. 3.8(*b*), and is much more complicated – albeit more accurate. All the circuit element values are now functions of the mouth radius.

Compared with the lips and the nostrils, the loading effect of the glottis is quite small. However, since the effect of glottal loading is part of a much more complicated problem – that of characterising the vocal cord system – a detailed discussion will be taken up in a separate section.

3.6 Glottal excitation

It has been suggested in the foregoing, that vocal tract excitation by periodic vibration of the vocal cords can be modelled by considering the cords to act as a simple acoustic source with a linear, time-invariant, internal impedance. This is only approximately true and, though the approximation is widely used in speech synthesis, the details of the real situation are worth some consideration. Such an analysis will give some insight into the extent to which source/system separation is valid.

The vocal cords produce voiced sounds by chopping up the air-flow through the glottis into pulses. The actual mechanism which causes the vocal cords to vibrate will be included later in the analysis; initially the cords will be assumed to vibrate independently of pressure and flow variations in the glottal region. The mechanism to be discussed is thus one of a time-varying aperture $A_g(t)$, at the glottis. The objective is to arrive at a linear, time-invariant model of vocal cord action; however, to do this some rather gross simplifications have to be made.

Fig. 3.9(*a*) shows a schematic diagram of the glottal region. The vocal 'cords' are simply folds of tissue which form a slit of variable width W across the glottis. This acoustic system can be translated into more familiar electrical terms and is illustrated in Fig. 3.9(*b*). The lungs and trachea are represented by a large variable capacitance and a short transmission line. The glottis is characterised by a series connection of time-variable inductance L_g, and resistance R_g, connecting the trachea and the vocal tract – also represented by a transmission line.

Air (current) from the lungs (capacitance) flows through the system and out through the lips. The time variations of L_g and R_g impose their periodic variations on this steady flow. Since both L_g and R_g are flow dependent, and since they are both time-variables, the situation is both non-linear and time-variant. For time durations corresponding to several oscillations of the vocal cords, the lung capacitance may be considered to be a battery providing a constant d.c. voltage. This approximation is enhanced by the tendency of the lungs to contract to keep the sub-glottal pressure, P_s, at a steady value. The acoustic impedance of the trachea, a large diameter tube, is small compared with the glottal impedance, thus the lung–trachea–

Fig. 3.9. (*a*) Schematic diagram of glottis; q is length, g is height, and W is width of the glottal opening. (*b*) Transmission line equivalent circuit of lungs, trachea, glottis and vocal tract. (*c*) Equivalent circuit of glottis and vocal tract, assuming that lungs and trachea provide a constant *sub-glottal pressure* P_s. (*d*) Simplified diagram of glottis, assuming that the vocal tract has negligibly small input impedance.

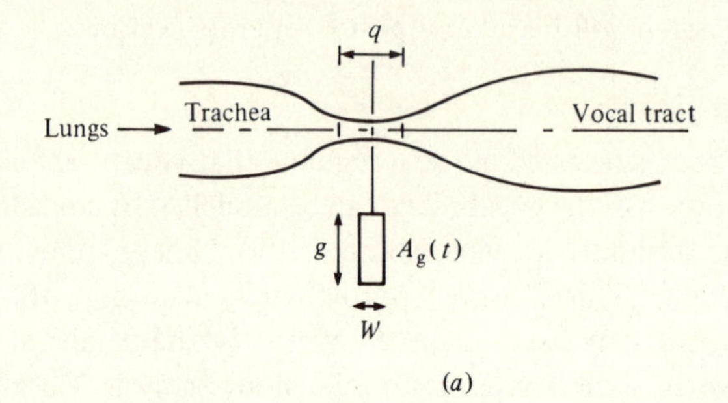

(*a*)

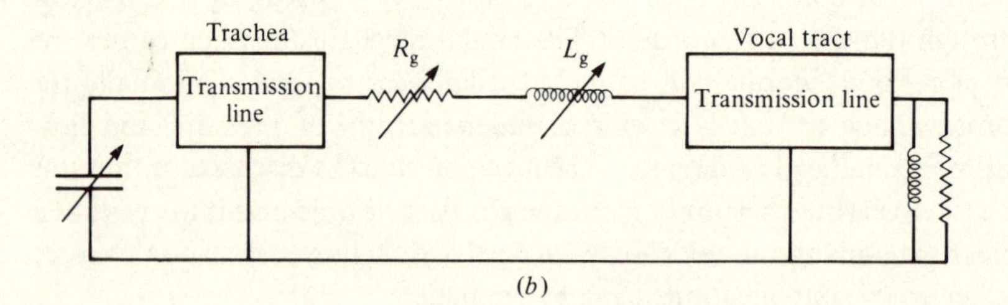

(*b*)

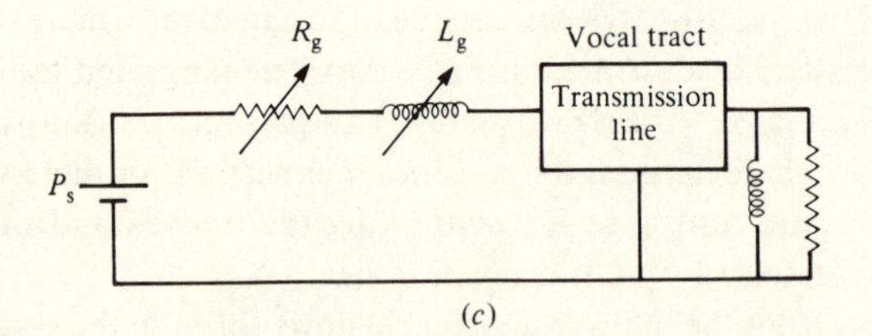

(*c*)

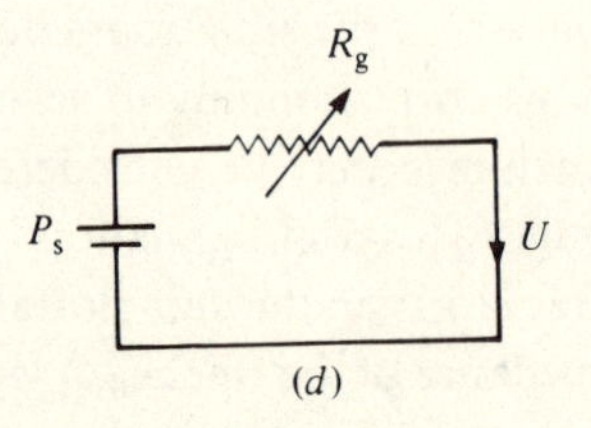

(*d*)

glottis system can be approximated by the network shown in Fig. 3.9(*c*). The size of the equivalent R_g and L_g of the glottis may be calculated by the usual techniques. The inductance is due primarily to the inertia of the air in the glottis. In terms of the relevant dimensions L_g can be shown to be

$$L_g = \frac{Dq}{A_g(t)}$$

where q is the length of the glottal region.

The equivalent resistance may be derived from either laminar or turbulent flow analysis, depending on the flow conditions. For large glottal opening laminar flow conditions apply, and the constriction resistance is essentially

$$R_l = \frac{12Dqg^2}{A_g^3(t)}$$

When turbulent flow sets in the resistance becomes

$$R_t = \frac{DUq}{2A_g^2(t)}$$

Flanagan (1972) suggests that a reasonable average value would be a linear combination of these two terms. He goes on to show that, for normal voicing, turbulent flow takes place for 80% of the time, and that $R_g = R_t$ is a good approximation.

The situation is very much further complicated by the time variation of L_g and R_g. In particular, current flow through a time-varying inductance is difficult to analyse. Flanagan (1972) has shown that the time constant L_g/R_g of the glottal impedance is small compared with one cycle of glottal oscillation. Thus it is not too inaccurate an approximation to consider the pressure drop across L_g to be negligible compared to that across R_g. If this is done then the time variable can be considered to act independently via the glottal area function $A_g(t)$. That is, the time rate of change of flow through L_g will be ignored.

The final approximation is to assume that the input impedance of the vocal tract is small compared with R_g. Thus the pressure/flow equation for the glottis can be obtained from the circuit of Fig. 3.9(*d*) as simply $P_s = UR_g$. Considering only the turbulent term

$$P_s = \frac{U^2 D}{2A_g^2(t)}$$

giving

$$U = A_g(t)\left(\frac{2P_s}{D}\right)^{1/2}$$

This relationship is now only time-varying through the glottal area function $A_g(t)$, which is assumed to act independently of the pressure and flow variations. However, the system is still non-linear, and the relationship between U, P_s and $A_g(t)$ is difficult to visualise. Essentially, P_s is held substantially constant, while $A_g(t)$ varies causing variations in U. Fig. 3.10 attempts to show these variables in a three-dimensional diagram. In order to derive a small-signal equivalent source, it is convenient to characterise $A_g(t)$ by an average value A_0 plus a varying component ΔA. Similarly, U can be written as a steady value U_0 plus a variation ΔU. The equation can now be written

$$U_0 + \Delta U = (A_0 + \Delta A)\left(\frac{2P_s}{D}\right)^{1/2}$$

giving

$$\Delta U = \Delta A\left(\frac{2P_s}{D}\right)^{1/2}$$

The term ΔU is the value of the small-signal equivalent current source, and is directly proportional to ΔA, the variation in glottal area. This signal source will act to force current into the vocal tract, and its internal impedance will be the slope of the three-dimensional surface parallel to the U and P_s axes, at A_0 and U_0, in the diagram of Fig. 3.10. This is an 'a.c.'

Fig. 3.10 Three-dimensional representation of the non-linear relationship between sub-glottal pressure P_s, glottal air flow U, and glottal area A_g.

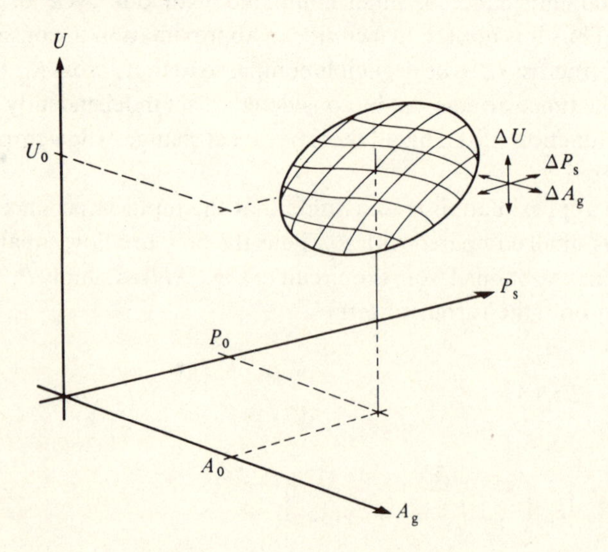

slope resistance, and is given by $r_g = \Delta P_s/\Delta U$ at $U = U_0$, and $A = A_0$, that is $r_g = U_0 D/A_0^2$.

It is traditional in speech synthesis to consider this small-signal current source to be separable from the vocal tract. This is mainly a matter of convenience, but the extent to which the approximation is valid can be assessed by comparing the input impedance of the vocal tract with the glottal source resistance r_g. Over most of the frequency range of interest, r_g is much larger, and can therefore be neglected. However, in the low-frequency region of the first formant, the tract input impedance is of the same order as r_g; thus, r_g may be assumed to damp the first formant in a non-linear manner. In vocal tract models, this interaction occurs naturally if a suitable glottal source resistance is included. But in terminal analogue synthesisers, the effects of such interaction must be incorporated in the control parameters, if indeed it is to be included at all.

This small-signal equivalent source is, of course, the result of a gross simplification. What happens in reality is that reflections of pressure and flow from the vocal tract will alter the conditions in the glottal region to such an extent that vocal cord oscillation will be affected. This kind of interaction can only be modelled by a full vocal cord simulation, a discussion of which will be continued in the next section.

3.7 Vocal cord model

In estimating the parameters of the vocal cord oscillatory system in terms of a linear, time-invariant current source, several sweeping simplifications had to be made. This was necessary to force what is rather complex behaviour to fit a somewhat simple model. One such assumption was that the vocal cords were, in effect, 'driven' by an external oscillator, and that the glottal area function $A_g(t)$ was unaffected by pressure and flow variations in the glottal region. In reality, the vocal cords and the glottal tract act together as an acoustic oscillator, and it is the interactions of pressure, flow and glottal dynamics that actually cause the oscillations.

Vocal cord oscillation is initiated by raising the sub-glottal pressure P_s to a significant value, and simultaneously increasing the muscular tension on the vocal cords. The pressure forces the cords apart, allowing flow to take place, thus releasing the pressure, which permits the cords to close again. This sequence repeats if the pressure and tension are maintained at correct values. The behaviour of such a system has been studied extensively by Flanagan and others over a number of years.

In an initial study, Flanagan & Landgraf (1968) modelled each vocal fold as a single lumped mass with compliance and non-linear resistance, moving in a plane perpendicular to the glottal axis. Fig. 3.11 shows the model, in which the two controlling parameters are P_s, the sub-glottal pressure, and Q, the vocal cord tension. Using this simple model of the vocal cords to drive a transmission line model of the vocal tract, it was shown that self-oscillation occurred for certain values of P_s and Q. The model behaved in a similar manner to the human system; frequency of oscillation proved to be substantially independent of the vocal tract configuration, and controllable by variation of P_s and Q. However, the model would not sustain stable oscillations at frequencies above that of the first formant of the vocal tract.

In later studies, Ishizaka & Flanagan (1972) and Flanagan & Ishizaka (1978) extended the model to include two-mass movement of the cords, movement along the axis of the glottal tract, and volume displacement in the glottal region. This two-mass model is shown in Fig. 3.12. Each vocal cord is represented by two masses coupled together. Again, realistic oscillations occurred when the model was connected to the vocal tract model. By varying P_s and Q, the model could be made to oscillate over a wide range of frequencies. Fig. 3.13 shows voicing frequencies F_0 against Q

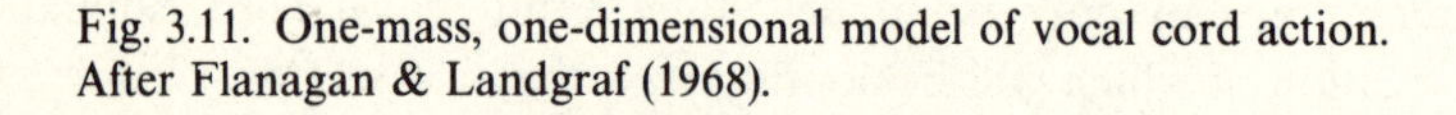

Fig. 3.11. One-mass, one-dimensional model of vocal cord action. After Flanagan & Landgraf (1968).

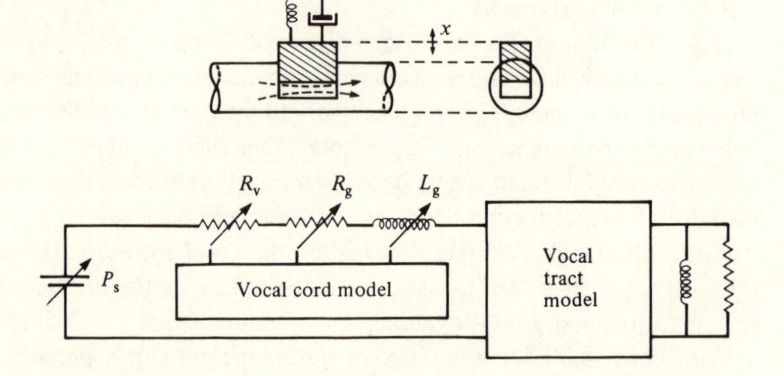

Fig. 3.12. Two-mass, two-dimensional model of vocal cord action. After Ishizaka & Flanagan (1972).

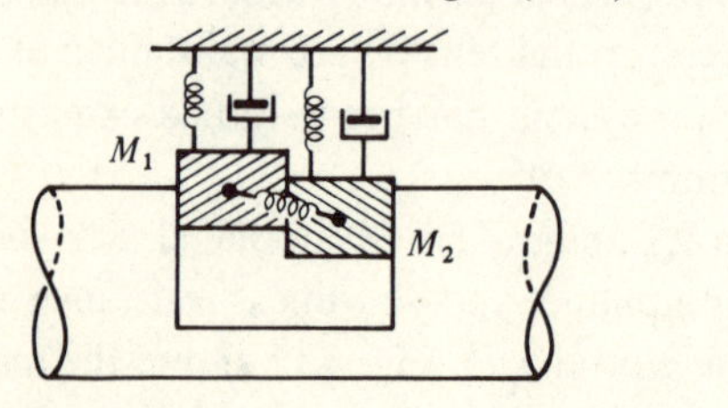

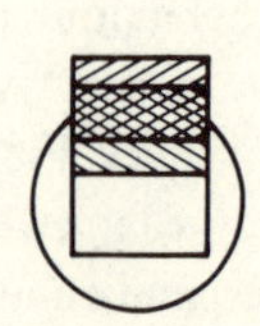

for constant P_s, and Fig. 3.14 shows F_0 against P_s for constant Q. In this model the pitch frequency F_0 could be made to go above the frequency of the first formant, giving a 'jump' effect, similar to characteristics of the human voice. Furthermore, the movement of the two masses corresponded well with the movements of the upper and lower edges of the vocal cords, as revealed by motion pictures of the human larynx.

Flanagan, Ishizaka & Shipley (1975) have shown that many of the physiological effects displayed by the more advanced models are not perceptually relevant, and conclude that for the purposes of speech synthesis a simple two-mass model in one dimension is quite adequate.

Fig. 3.13. Variation of fundamental frequency F_0, against cord tension Q, for constant sub-glottal pressure P_s. After Ishizaka; Flanagan (1972).

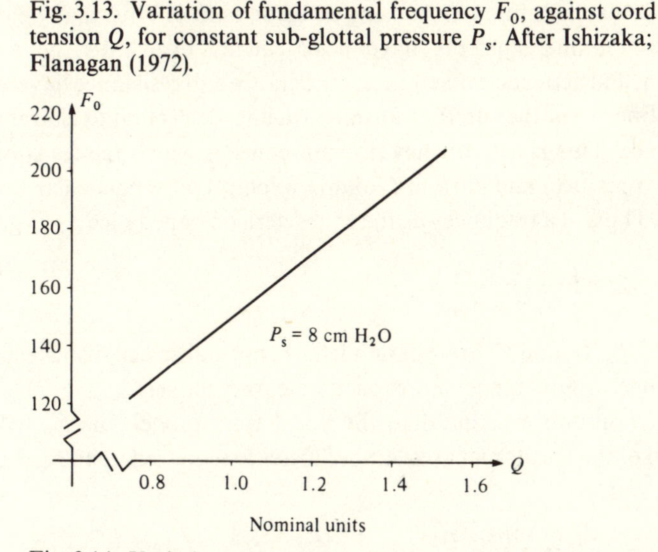

Fig. 3.14. Variation of fundamental frequency F_0, against sub-glottal pressure P_s, for constant cord tension Q. After Ishizaka & Flanagan (1972).

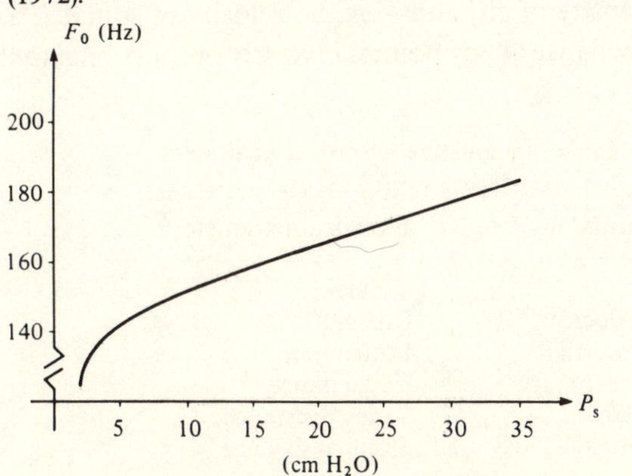

Finally, Flanagan & Ishizaka (1976) have shown that this vocal cord model connected to a transmission line model of the vocal tract can give very realistic speech sounds, and is capable of reproducing many natural vocal effects.

3.8 Electrical transmission line analogues

It has been shown that plane wave sound propagation in an acoustic tube is analogous to the propagation of electric waves along a transmission line. The analogous relationships are summarised in the table of Fig. 3.15.

This analogy is not only useful in making electrical models of the vocal tract, it is also helpful in visualising acoustic events. Thus many electrical transmission line concepts have been carried over into the acoustic situation, and acoustic capacitance, inductance and resistance have already been defined. Another useful transmission line idea is that of characteristic impedance. This is a frequency domain concept which relates sinusoidal voltage (pressure) and current (volume velocity), at a particular point on the line. Thus, at a frequency ω, the characteristic impedance Z_a, is given by

$$Z_a = \left(\frac{R_a + j\omega L_a}{G_a + j\omega C_a}\right)^{1/2}$$

where R_a, L_a, G_a and C_a are, per unit length, normalised acoustic resistance, inductance, conductance and capacitance, respectively.

For a non-uniform line, like the vocal tract model, the Z_a will be a function of the vocal tract area $A(x)$. If losses are ignored, then $R_a = G_a = 0$, and

$$Z_a = \left(\frac{L_a}{C_a}\right)^{1/2} = \frac{DV}{A(x)} \qquad \text{(see section 3.2)}$$

where D is the density of air, and V is the velocity of sound. That is, the characteristic impedance at any point is inversely proportional to the cross-sectional area.

Fig. 3.15. Table of acoustical/electrical analogues.

Acoustical parameter	Electrical parameter
Pressure	Voltage
Volume velocity	Current
Air mass inertia	Inductance
Air compressibility	Capacitance
Viscous loss	Series resistance
Heat conduction loss	Shunt resistance

Another useful transmission line concept is that of propagation constant P_a, defined as

$$P_a = [(R_a + j\omega L_a)(G_a + j\omega C_a)]^{1/2}$$

The significance of P_a is that it characterises the propagation effect. In general P_a will be a complex quantity, $P_a = a + jb$; thus at a point x, along the line from a sinusoidal source $e^{j\omega t}$, the wave will be

$$e^{P_a x}e^{j\omega t} = e^{(a+jb)x}e^{j\omega t} = e^{ax}e^{j(bx+\omega t)}$$

The term e^{ax} is an attenuation factor, and the e^{jbx} term is a phase shift. For a lossless line,

$$P_a = j\omega(L_a C_a)^{1/2}$$

so that there is no attenuation. The phase shift is therefore $\omega(L_a C_a)^{1/2}$ radians per unit length, which corresponds to a time delay of $(L_a C_a)^{1/2}$ seconds per unit length.

In modelling the vocal tract it is not possible to use a real transmission line with distributed elements. Instead, a lumped component model is used, in which small segments of the line are approximated by inductor–capacitor–resistor networks. For a very small segment length Δx, a transmission line segment can be represented by the Pi or Tee network shown in Fig. 3.16. Approximately, the required inductance, capacitance, resistance and conductance of the segment will be given by $L_a\,\Delta x$, $R_a\,\Delta x$, $C_a\,\Delta x$ and $G_a\,\Delta x$, respectively, where L_a, C_a, R_a, G_a are the per unit length constants of the line at the point under consideration. Many such segments may be connected together to give the required length of line to simulate a vocal tract. A simple case is when the acoustic tube is assumed lossless, that is when $R_a = G_a = 0$. Then the characteristic impedance of the transmission line model becomes $Z_a = DV/A(x)$. Similarly, the propagation constant is $P_a = j\omega(L_a C_a)^{1/2}$ and the time delay for each segment is given by $\Delta T = \Delta x(L_a C_a)^{1/2}$. Since $L_a = D/A(x)$, and $C_a = A(x)/DV$, the delay $\Delta T = \Delta x/V$. This confirms the validity of the model, since V, the velocity of the sound wave in the acoustic tube is, by definition, $\Delta x/\Delta T$.

Fig. 3.16 Pi and Tee sections of a discrete electrical transmission line.

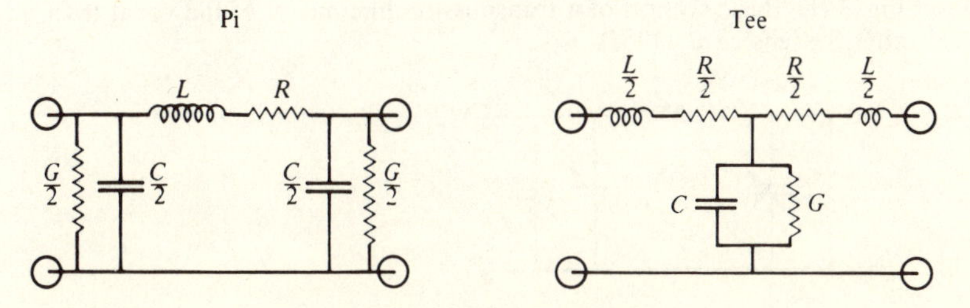

Dunn (1950) constructed a lossless transmission line model of this kind, consisting of 25 segments, each representing 0.5 cm of acoustic tube with cross-sectional area of 6 cm^2. He showed that reasonable vowel sounds could be produced by simply introducing a large inductance in series with the line at appropriate points, to simulate vocal tract narrowing due to the tongue hump. The arrangement is illustrated in Fig. 3.17.

A more sophisticated transmission line analogue was devised by Stevens *et al.* (1953), who used 35 sections with variable L_a and C_a elements, so that 'area' could be varied between 0.17 and 17 cm^2 in ten steps. In addition each section had a variable shunt resistance to permit losses to be taken into account. Fig. 3.18 shows a typical section. Using area data taken from X-ray photographs, this model produced very convincing vowel sounds with realistic formant frequencies and amplitudes. To simulate fricative sounds, random noise was injected into the line at constriction points. Since each section had to be carefully set up by hand, it was not possible to produce variable speech sounds with this transmission line model.

An attempt to make a transmission line model which was dynamically variable was reported by Rosen (1956). This model used 13 sections of variable inductor–capacitor circuits. Inductance variation over a 250:1 range was achieved using saturating reactance, and variable capacitance over a similar range by using feedback amplifiers with variable gain. This analogue was able to synthesise realistic diphthongs, fricatives and some stops. The limitation to this model was the difficulty with which control over L_a and C_a was implemented, and the problem of specifying and controlling losses in the tract.

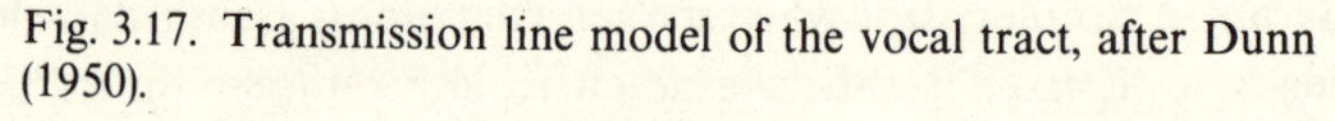

Fig. 3.17. Transmission line model of the vocal tract, after Dunn (1950).

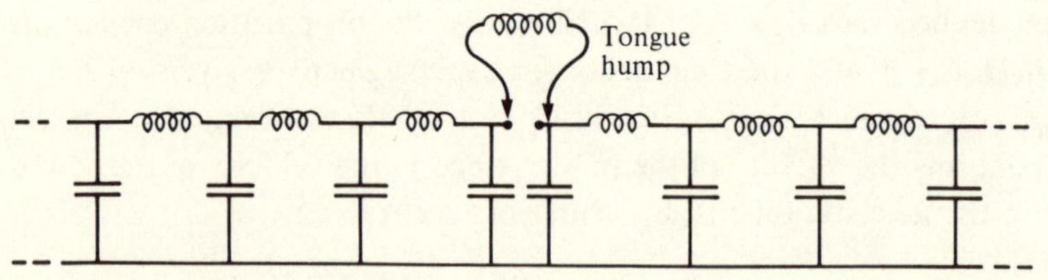

Fig. 3.18. Basic section of a transmission line model of the vocal tract, after Stevens *et al.* (1953).

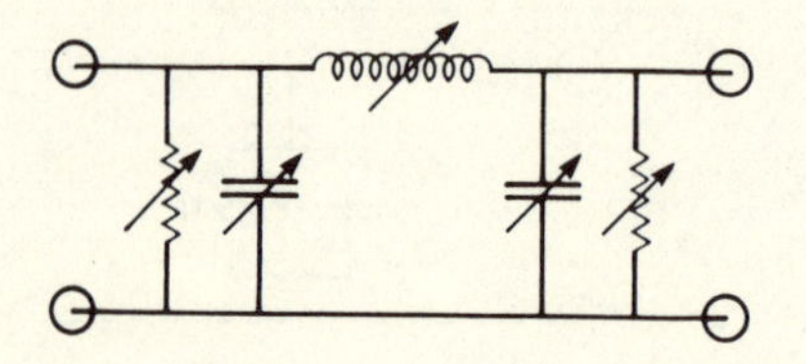

A more recent attempt (Keltai, 1980) to make a dynamically variable transmission line analogue, using variable electronic transformers between fixed inductor–capacitor Pi sections, met with similar problems and success.

3.9 Computational model

The two basic equations for the lossless acoustic tube were shown in section 3.2 to be

$$\Delta P = \frac{D}{A(x)}\frac{\partial U}{\partial t}\Delta x \qquad \text{and} \qquad \Delta U = \frac{A(x)}{DV^2}\frac{\partial P}{\partial t}\Delta x$$

As the increment of length Δx goes to zero, these expressions may be written in partial differential form,

$$\frac{\partial P}{\partial x} = \frac{D}{A(x)}\frac{\partial U}{\partial t} \qquad \text{and} \qquad \frac{\partial U}{\partial x} = \frac{A(x)}{DV^2}\frac{\partial P}{\partial t}$$

By taking second differentials, these equations can be combined to give

$$\frac{1}{V^2}\frac{\partial^2 P}{\partial t^2} = \frac{1}{A(x)}\frac{\partial}{\partial x}\left[A(x)\frac{\partial P}{\partial x}\right]$$

This is known as Webster's horn equation, which can only be solved if $A(x)$ is a well behaved analytic function. Unfortunately, in the acoustic tube model of the vocal tract, $A(x)$ is not known as a mathematical function. That is, $A(x)$ is only given as a set of samples at discrete points along the length of the tract. Since $A(x)$ is numerical, a numerical solution to Webster's equation must be found. This is done by approximating the spacial differential equation by a difference equation. Thus the continuous variable x is replaced by a discrete variable $k\,\Delta x$, and $A(x)$, $P(x)$, $U(x)$ become discrete variables $A(k\,\Delta x)$, $P(k\,\Delta x)$, $U(k\,\Delta x)$, or more concisely $A(k)$, $P(k)$, $U(k)$, where k is an integer. The first derivative

$$\frac{\partial P}{\partial x} \quad \text{is approximately} \quad \frac{P(k) - P(k-1)}{\Delta x}$$

and the second differential

$$\frac{\partial^2 P}{\partial x^2} \quad \text{is approximately} \quad \frac{P(k+1) - 2P(k) - P(k-1)}{\Delta x^2}$$

$P(k)$ is still a continuous time variable; so that, assuming solutions to be of the form

$$P(k) = \bar{P}(k)e^{j\omega t + jbx}$$

a recurrence relationship can be established between $P(k)$ and $P(k+1)$. Hence, given suitable boundary conditions, a numerical solution can be easily computed, a 'solution' in this case being a set of $P(k)$ for any specific value of the frequency variable ω.

Coker (1968) has used this technique with lossless boundaries (mouth as a short-circuit, glottis as an open circuit) to investigate the resonant frequencies of the vocal tract for particular articulatory configurations. Since Webster's equation is itself lossless, this method gives resonances with zero bandwidth. Because of the absence of loss in the system this computational model cannot be used to synthesise speech directly. Instead, it is used as an intermediate step to go from vocal tract area functions to formant frequencies. The resulting formant trajectories can then be used to drive a formant synthesiser, in which formant bandwidths are set at realistic values.

Another computational model is the computer analysis of an electrical transmission line analogue. If the line is lossless then the analysis is effectively the same as a discrete time solution of Webster's equation. However, if losses are included then a more realistic and difficult situation ensues. Modern network analysis programs can easily cope with the size of the network involved, in either the frequency or the time domain. But such general purpose programs are not really suitable, since they are not optimised for this particular problem.

Flanagan *et al.* (1975) have solved the problem of simulating the lossy transmission line by setting up difference equations for each section of the line. The equations are solved at discrete time intervals to give samples of pressure and flow at each point in the line. The values of pressure and flow at one time instant are used to determine the losses for the next time interval. Thus the losses are controlled and updated in a realistic manner. In particular, the onset of turbulent flow is detected, and a noise source with internal resistance of appropriate values is inserted at the point of constriction. Fig. 3.19 shows a typical section with the optional noise source included. The model uses 20 such sections, and sound output via wall

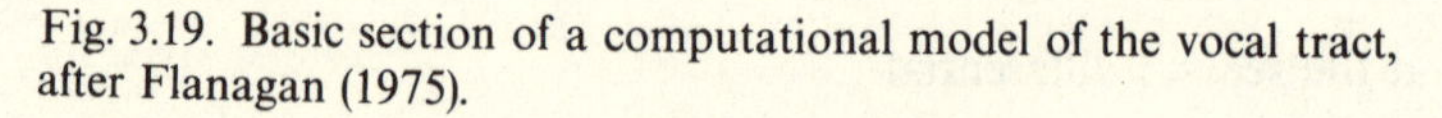
Fig. 3.19. Basic section of a computational model of the vocal tract, after Flanagan (1975).

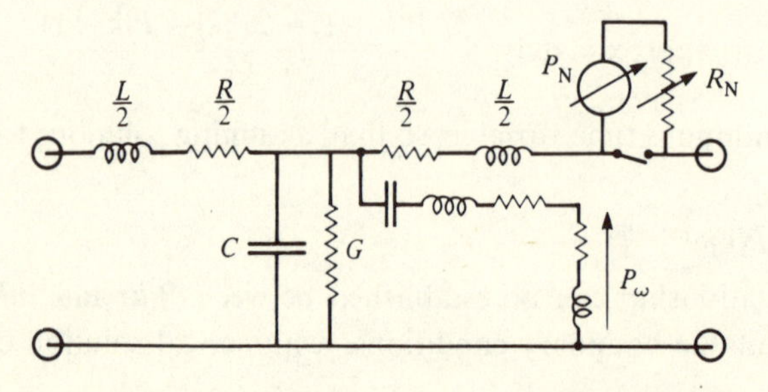

vibration is taken into account by summing the P_ω components from each section and adding them into the output. This computational model is the most complete simulation of the vocal tract reported in the literature.

3.10 Digital transmission lines

The discrete (or digital) transmission line makes use of another classical transmission line concept, that of travelling waves. The pressure (voltage) distribution along a transmission line is visualised as consisting of two travelling waves, propagating in opposite directions. The pressure P at any point on the line is considered to be made up of two components, a forward wave P^+, and a backward wave P^-. Thus $P = P^+ + P^-$. This concept not only simplifies the calculations but also gives insight into the behaviour of the line at impedance discontinuities.

At the junction of two sections of line with different characteristic impedances, each wave is visualised as undergoing propagation and reflection. The forward wave P^+ is partly propagated and partly reflected, the reflected fraction rP^+ now becomes part of the backward wave, and the propagated part continues down the line. The factor r is called the reflection coefficient, and is defined in classic transmission line theory as

$$r = \frac{Z_f - Z_b}{Z_f + Z_b}$$

where Z_f is the characteristic impedance forward of the junction, and Z_b is the characteristic impedance of the line backward of the junction. Note that $-1 \leqslant r \leqslant +1$.

At the junction, P^+ is continuous, hence the propagated fraction must be $(1+r)P^+$, so that $P^+ = (1+r)P^+ - rP^+$.

Exactly similar considerations apply to the backward wave P^-, except that since it travels in the opposite direction Z_f and Z_b are reversed. Thus the reflection coefficient to the backward wave is

$$\frac{Z_b - Z_f}{Z_b + Z_f} = -r$$

The propagated part of P^- is therefore $(1-r)P^-$, and the reflected part is $-rP^-$. In this case the reflected fraction adds to the forward going wave.

This situation is depicted in Fig. 3.20, which shows a single discontinuity in an acoustic tube. In the lossless case the acoustic impedances Z_f and Z_b are simple functions of the cross-sectional area $A(x)$. Thus $Z_f = DV/A_f$ and $Z_b = DV/A_b$, hence $r = (A_b - A_f)/(A_b + A_f)$.

To use this analysis in a situation where there are multiple discontinuities, and therefore multiple reflections, it is necessary to discretise both time and tube length. Consider the segmented acoustic tube shown in Fig. 3.20. The P^+ wave travelling from left to right will propagate undisturbed from point one until it reaches the second discontinuity at point two. Similarly, the backward wave will propagate from point three without reflecting until it reaches point two. The time taken for P^+ to go from point one to point two, and for P^- to go from three to two, is simply $\Delta T = \Delta x/V$, where V is the velocity of sound and Δx is the length of a tube segment.

Thus the pressure variations at point two (and every other point), need only be calculated at discrete time intervals ΔT. If these intervals are small enough then the resulting sampled data sequence can represent all the information required. The criterion for determining the sampling interval ΔT is given by the Nyquist sampling theorem; this requires ΔT to be less than or equal to $1/2\,B$, where B is the bandwidth of signal to be sampled. Thus the segment length Δx of an acoustic tube model is related to the rate $1/\Delta T$, at which samples of pressure and flow are calculated.

For synthetic speech with a bandwidth of 4 kHz, a sampling rate of 8 kHz is a minimum. This gives a ΔT of 0.125 ms and a segment length Δx of 4.325 cm. For a typical vocal tract length of 17.5 cm this necessitates a minimum of four sections. From a space sampling point of view, the same result ensues. At 4 kHz, the wavelength of the sound in the tube is 8.75 cm. According to Nyquist, a sine wave needs to be sampled at least twice within

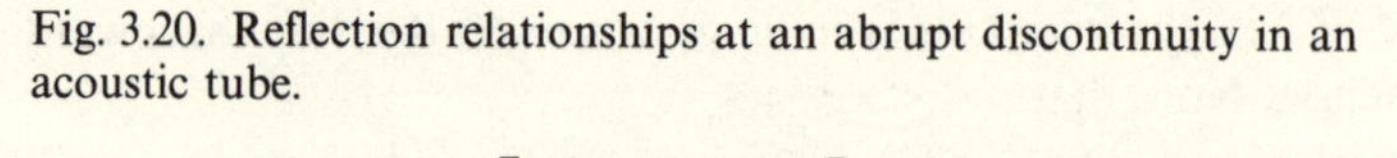

Fig. 3.20. Reflection relationships at an abrupt discontinuity in an acoustic tube.

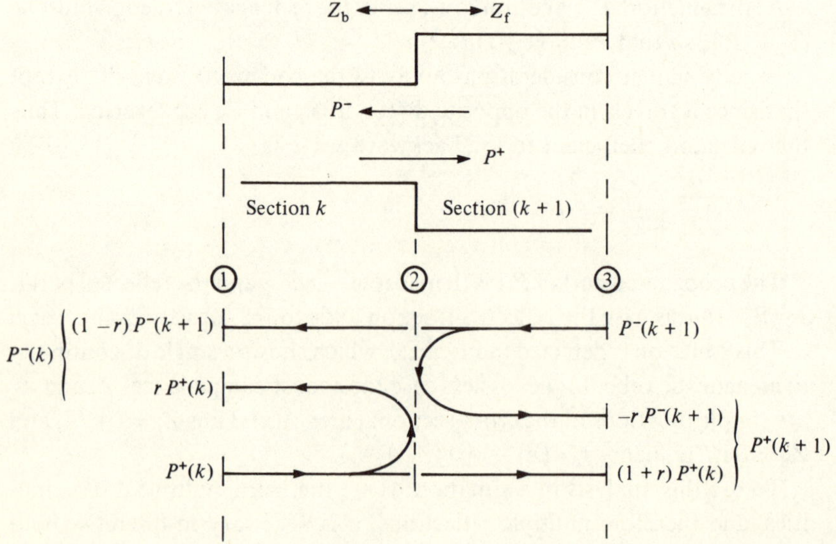

each cycle. Thus Δx would have to be $8.75/2 = 4.375$ cm. A four-segment acoustic tube model of the vocal tract is shown in Fig. 3.21.

The reflection relationships at the junction of two segments of tube are depicted in Fig. 3.22. This figure may be interpreted as being a flow graph for a computer program, or a block diagram for an analogue or digital circuit. The boxes labelled $\Delta T/2$ are time delays which represent the transit times between discontinuities.

Representation of the vocal tract by only four segments of acoustic tube is really too inaccurate to merit the name of vocal tract model. To invoke Nyquist again, the area function $A(x)$ must be sampled more frequently if the variations in its shape are not to be lost. It is, of course, possible to sample $A(x)$ at any convenient interval Δx, and have as many segments as required. However, this results in an increase not only in the number of sections, but in the time sampling rate as well. Thus if M is the number of sections, and L is the total tract length, then $\Delta x = L/M$; so that the time

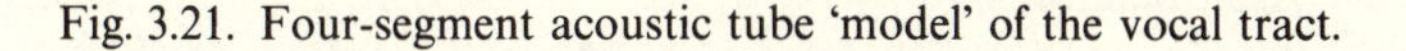

Fig. 3.21. Four-segment acoustic tube 'model' of the vocal tract.

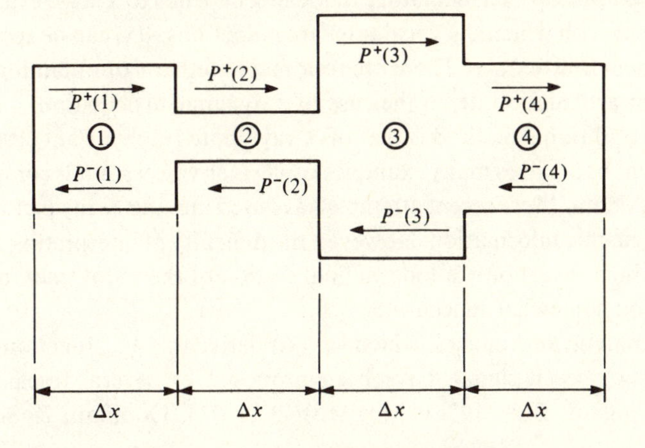

Fig. 3.22. Computational diagram of reflection relationships for a digital line model of the vocal tract.

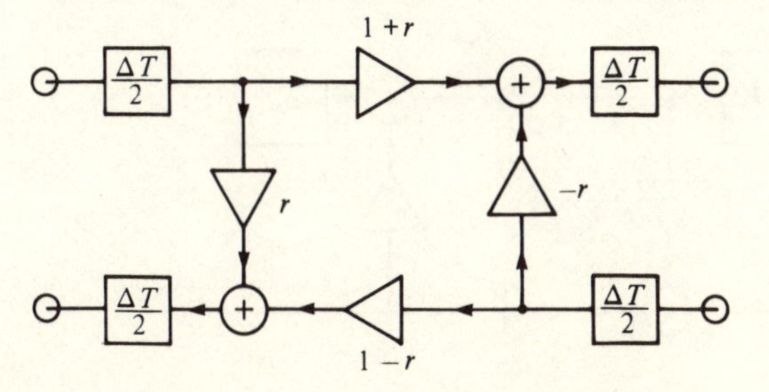

sampling period is $\Delta T = \Delta x/V = L/MV$. Hence the number of reflection calculations per second will be $M/\Delta T = VM^2/L$. This square law relationship means that doubling the number of sections quadruples the computational load. The situation can be improved by a factor of two if the flow graph of Fig. 3.22 is changed to that of Fig. 3.23. The system is now no longer a true model, since a time delay has been added to the computation. But the section delay is now $2\Delta T$, so that the time sampling rate is effectively halved.

This formulation was first given by Kelly and Lochbaum (1962), who used a sampling rate of 20 kHz with a 21-section tube model of the vocal tract. The glottis was represented by a tube of small ($0.2\,\text{cm}^2$) fixed area, and the mouth by zero acoustic impedance, giving a reflection coefficient $r = -1$. Losses were introduced into the system at each discontinuity by making the difference between the reflection coefficient and the propagation coefficient slightly less than unity.

3.11 Area functions

The success of articulatory modelling depends to a large extent on the accuracy with which the vocal tract area functions $A(x)$ can be specified for a particular utterance. There are four main methods of obtaining $A(x)$ for a given articulation or, in the case of a dynamic articulation, a set of $A(x)$. The traditional method is that of X-ray photography. Fant (1960) in his excellent book gives many examples of $A(x)$ for vowel sounds computed from X-ray data. More recent attempts have used moving X-ray pictures to capture dynamic information. However, the difficulty of interpreting X-ray images, which reveal only a longitudinal section of the vocal tract, makes this method somewhat inaccurate.

More analytic approaches, which seek to derive the $A(x)$ function from the acoustic speech signal, have been proposed by several researchers, notably Paige & Zue (1970), Mermelstein (1971), Gopinath & Sondhi

Fig. 3.23. Alternative computational diagram of the digital transmission line model.

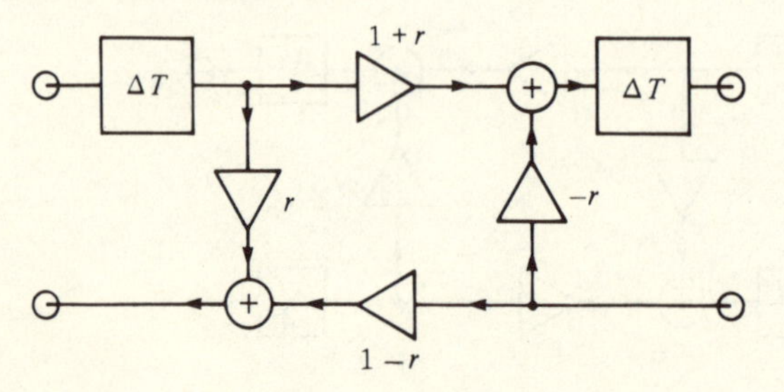

(1970), Stansfield (1973) and Wakita (1973). The basic concept behind these approaches is that, if the transfer function of the vocal tract is known, then the $A(x)$ can be derived uniquely. However, finding the transfer function of the tract from the acoustic signal is not a trivial problem. A knowledge of the formant frequencies can be used to infer an appropriate function, but again finding formant frequencies is not an easy task. Only in the case of vowels do these methods seem to work at all well.

Perhaps the most interesting approach to deriving appropriate sets of $A(x)$ is that proposed by Coker (1976). This is a synthesis-from-text method which has, as input, a phonetic description of the required utterance. The area function $A(x)$ is derived by a linear transformation of the positions of the basic articulators, the tongue tip, lips, velum, glottis and tongue body. Coker has shown that these primary variables have the minimum number of degrees of freedom to permit realistic speech to be synthesised. That is, $A(x)$ is constrained, by the variables and the transformation, to those shapes which result in speech-like sounds. These primary variables are not arbitrary, but are inferred from a study of articulatory movements. They are confirmed by a principal axis analysis, which also gives the precise transformation required to convert them into an $A(x)$ function.

The primary variables themselves are generated from phonetic text by recourse to look-up tables and a set of context rules. An important feature of this procedure is the smoothing of the variables by time constants

Fig. 3.24. Conceptual diagram of a technique for finding articulatory parameters, using a spectral match with real speech.

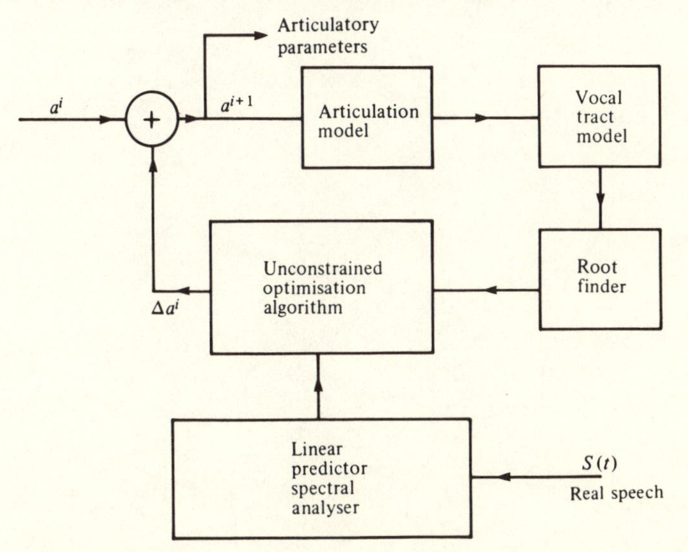

appropriate to physiological constraints on the articulators. This gives the effect of co-articulation between the phonetic elements of the utterance. The success of this strategy is confirmed by the high quality of the speech synthesised.

Finally, a more recent attempt to derive area functions from real speech has been reported by Levinson, Coker & Schmidt (1981). This uses the model proposed by Coker, but without the phoneme look-up table. Instead, the primary variables are adjusted, by an unconstrained optimisation technique, so that the formants of the synthetic speech match those of a given utterance. Fig. 3.24 shows the arrangement. The input speech is analysed into 10 ms spectral frames and the vocal tract model is adjusted to give similar formants on a frame by frame basis. The starting point for the optimisation algorithm at a particular frame is the tract configuration already derived for the previous frame. This ensures continuity of the $A(x)$ function. The resulting set of $A(x)$ functions can then be used to resynthesise the utterance, which, as might be expected, is very close to the original.

The great virtue of this method is that it can be used to derive articulatory control signals from real speech. An analysis of a sufficiently large body of such signals ought to give valuable clues to the basic rules of articulation of a particular language. This in turn could lead to more efficient ways of synthesising speech from phonetic input.

4 *Spectrum synthesis*

Modelling speech by modelling its spectrum

'The complete formation of synthetic speech... requires control of the three speech properties, pitch, spectrum and intensity, as well as a means of switching from one sound stream to the other.'
H. Dudley, 1939a

4.1 Introduction

In the previous chapter it was noted that the strategy of articulatory synthesis is to model the human vocal tract in as much physiological detail as possible, under the assumption that similar systems will give similar signals. The alternative approach, discussed here, is to produce synthetic speech directly by modelling the speech signal itself, rather than the system in which it originates. This has the advantage that the effects of obscure articulatory events, difficult to measure and model, can be readily reproduced at the acoustic level, by direct copying. Furthermore, acoustic details of the speech signal which are not perceptually relevant can quite conveniently be omitted from the modelling process.

Theoretically, at least, a good articulatory ('system') model should give better quality speech from a lower control data rate, since co-articulation effects are built into the model dynamics. In contrast, the 'signal' model is more difficult to control because of the need to copy (or account for by rules) every perceptible nuance of the speech signal. That signal-model synthesisers are more commercially successful than articulatory models may be accounted for by two facts. The first is that signal models are easier and cheaper to construct. The second is that, despite the ostensibly more difficult control problem, the speech signal is so readily accessible and measurable that it has proved more amenable to analysis than articulatory movements which, by their very nature, are difficult to observe and quantify.

Signal models of speech may operate in either the time or the frequency domain. Since speech is more perceptually relevant in terms of its short-time spectra, rather than its time waveform, a spectral model is particularly convenient. The short-time spectra of speech, as exemplified by spectrograms, have definite structural characteristics that are not too difficult to model. For voiced speech the spectra are harmonic at voice-pitch frequency, and are shaped by an 'envelope' having three to five resonant peaks (formants) and possibly an anti-resonance. This type of spectrum can be generated by exciting an appropriate resonant system with a periodic pulse train. For unvoiced speech, the spectra may be characterised as wideband noise, shaped by one or two resonant peaks and one anti-resonance. This may be generated quite simply by exciting an appropriate resonant system with white noise. Fig. 4.1 shows the minimal configuration for synthesising speech-like spectra.

The wave shape of the periodic pulse is not critical, since its spectrum can be shaped by the resonant system. Likewise, spectral colouring of the white noise source can be taken account of in the design of the resonant system. Control signals for such a basic arrangement are simply: F_0, the fundamental frequency of the pulse source; A_0, its amplitude; A_N, the amplitude of the noise source; and $\mathbf{V}_R$, a vector which specifies the transfer function of the resonant system. Depending on the complexity of the synthesiser, $\mathbf{V}_R$ may have between eight and twenty components. The variety of the spectrum modelling techniques described in the literature is due mainly to the variety of ways in which the resonant system may be

Fig. 4.1. Minimal configuration for synthesising speech-like spectra.

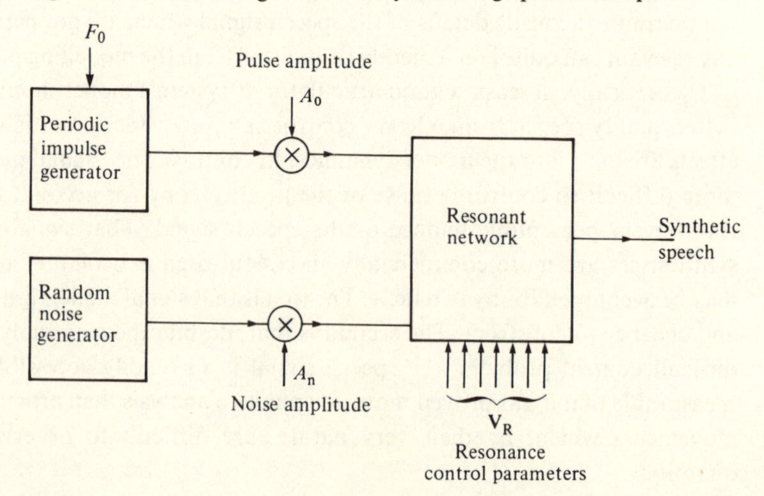

implemented and controlled. Almost always, periodic pulses and random noise are used as the raw material of the synthetic spectral shapes.

4.2 Channel synthesisers

The first true electronic speech synthesiser was invented by Dudley (1936, 1939*a*, *b*), and demonstrated at the New York world fair in 1939. This machine consisted essentially of ten bandpass resonators excited by pulse and noise sources. The outputs of the ten fixed frequency resonators were mixed in different proportions by potentiometers controlled by ten keys. Fig. 4.2 shows a schematic diagram of the machine which was adapted from a vocoder to facilitate manual control. The controls were all manual (hence ten keys) and designed to enable a skilled operator to produce intelligible speech in real time. The device was essentially an electronic equivalent of von Kempelen's speaking machine, and trained operators were able to astonish visitors to the world fair with demonstrations of synthetic speech.

This first speech synthesiser, called the 'Voder' by its inventor, developed out of work on vocoders (another Dudley invention). This was the case in almost all the early work on speech synthesisers until about 1970, when the possibility of using synthesisers as voice output devices for computers became a more important motivation. The manual control system of the Voder was an experimental necessity at that time, since digital computers were not yet available. Other early synthesisers (Cooper, Liberman & Borst, 1951; Lawrence, 1953) used optical scanning to record and replay control data.

In a channel synthesiser, the resonant network consists of several fixed-frequency, bandpass filters connected in parallel. This resonant system is excited by periodic pulses and/or random noise, and the output signal is a weighted sum of the filter outputs. More (or less) spectral detail can be

Fig. 4.2. Schematic diagram of the Voder. After Dudley (1939*a*).

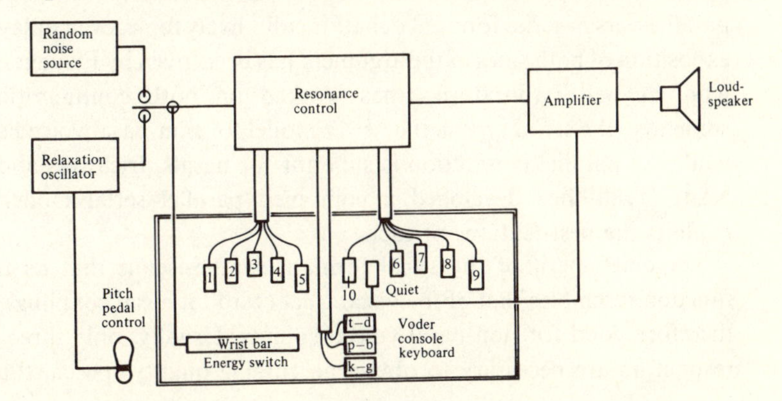

obtained by using more (or fewer) channels. High quality machines used up to thirty channels. Obviously, the more channels used the greater the amount of data required to effect control. Efforts to reduce the amount of control data, without loss of spectral detail, led to the development of the formant synthesiser – a much more efficient synthesis method, which is described in the next section.

4.3 Formant synthesisers

In the channel synthesiser, the three to five formant peaks of the typical speech spectrum are created by controlling the amplitudes of the channel filter outputs. The twenty or so channels must all be fed with control signals to specify their output amplitude. In the formant synthesiser, only three to five resonant circuits are required, but the resonant frequency of each resonator must be variable as well as the amplitude. This means that fewer control signals are required to specify a particular spectral shape. The exact shape of the spectrum may not be exactly as required, depending on the nature and arrangement of the resonant circuits, but careful design can ensure a close enough 'fit' to give excellent speech quality.

Formant synthesis appears to have grown out of two separate lines of research. The first, already mentioned, was the need to reduce the control data rate in channel vocoders. The obvious simplification in this context was to connect the variable resonant circuits in parallel, giving the 'parallel formant' synthesiser. The other approach was through the desire to construct an 'economy' vocal tract model, without bilateral circuits and with only the minimum number of resonators to produce speech-like sounds. The natural way to achieve this was to use a cascade of variable resonators, giving a 'serial formant' synthesiser.

These two different solutions to the formant synthesis problem have provoked much discussion in the literature as to which is the better, and the parallel-versus-serial formant debate is still a lively topic, even today. A fair exposition of both sides of the argument has been given by Flanagan (1957) who, at Bell Laboratories, has worked on both configurations. In summary, the serial type is the better model for non-nasal voiced sounds, while the parallel connection is superior for nasals, fricatives and stops. Klatt (1980) has developed a combined parallel–serial model which exploits the best features of each.

The chief merit of the serial formant synthesiser is that its transfer function resembles that of the vocal tract (without nasal coupling), and is therefore good for non-nasal voiced sounds. Usually, only three to five resonators are necessary to obtain acceptable quality speech, thus only three to five formant frequencies and bandwidths need to be specified to define the spectrum. The simplest form of resonant low pass filter has a function

$$\frac{\omega_0^2}{s^2 + Bs + \omega_0^2}$$

where B is the bandwidth, ω_0 is the resonant frequency and s is the complex frequency variable. Fant (1956) has shown that simple, low pass, second-order functions of this kind are quite adequate if a compensating network is used to correct for the absence of higher formants. This correcting function, illustrated in Fig. 4.3, depends on the number and frequency of the formants used, but in a practical situation can be approximated by a fixed network. The fact that formant amplitudes cannot be controlled individually does not matter if the bandwidths are adjusted correctly. Fant (1956) showed that the formant amplitudes are, in fact, predictable from a knowledge of the formant frequencies and bandwidths. In a practical synthesiser it is not too much of an approximation to fix the bandwidths of each formant so that only three to five formant frequencies are required to specify any non-nasal voiced sound. This makes the serial formant synthesiser very attractive in terms of control data economy. However, the model fits only non-nasal voiced speech, and stops, nasals and fricatives have to be catered for by extra resonators designed to introduce additional

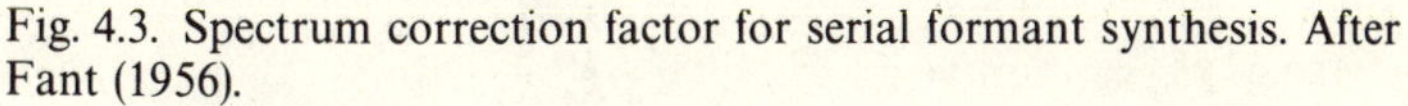

Fig. 4.3. Spectrum correction factor for serial formant synthesis. After Fant (1956).

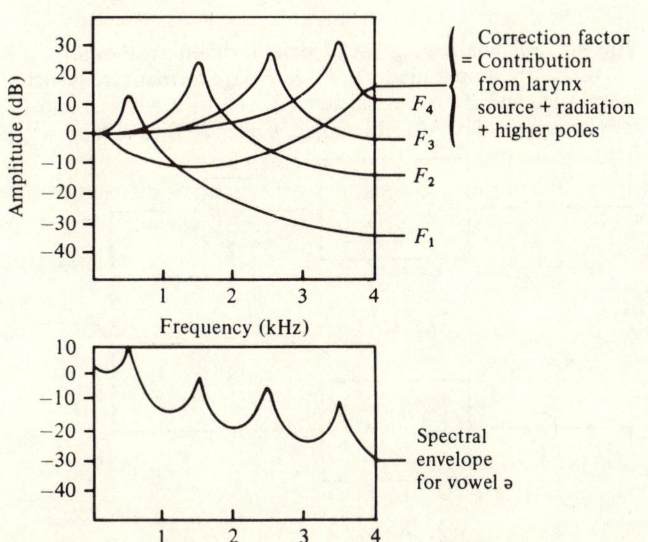

poles and zeros into the transfer function. A typical serial formant synthesiser, developed from Fant's work in this field is the OVE synthesiser (Fant, 1973), and is shown schematically in Fig. 4.4.

A variety of methods have been used to produce nasals, stops and fricatives in serial formant synthesisers (Flanagan & House, 1956; Stevens, Bastide & Smith, 1960; Flanagan, Coker & Bird, 1962; Rabiner, 1968*a*). As

Fig. 4.4. Schematic diagram of the OVE II synthesiser. F_0 is pitch; A_N, A_0, A_H & A_C are amplitudes; F_1, F_2 & F_3 are formant frequencies; K_0 is a fricative zero and K_1, K_2 are fricative poles. After Fant (1973).

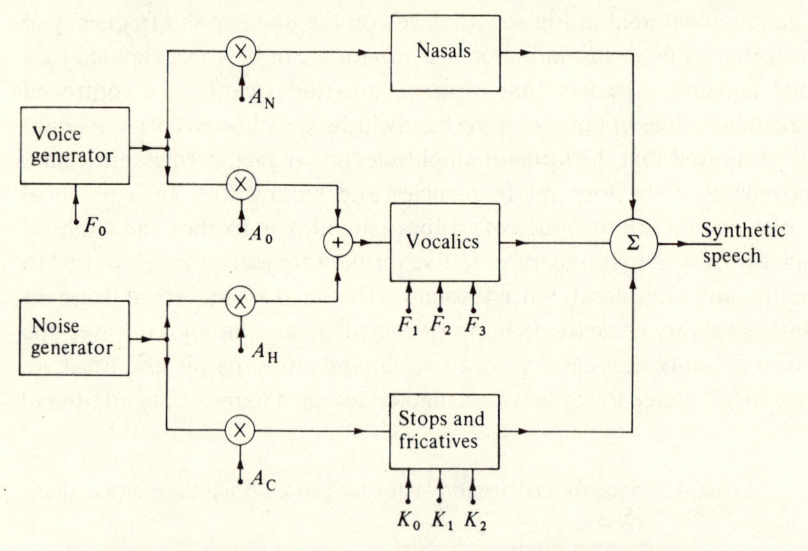

Fig. 4.5. Schematic diagram of serial formant synthesiser. F_0 is pitch; A_v & A_N are amplitudes; F_1, F_2 & F_3 are formant frequencies; P_N is a nasal pole and Z_N is a nasal zero; Z_f is a fricative zero and P_{F_1} & P_{F_2} are fricative poles. After Flanagan, Coker & Bird (1962).

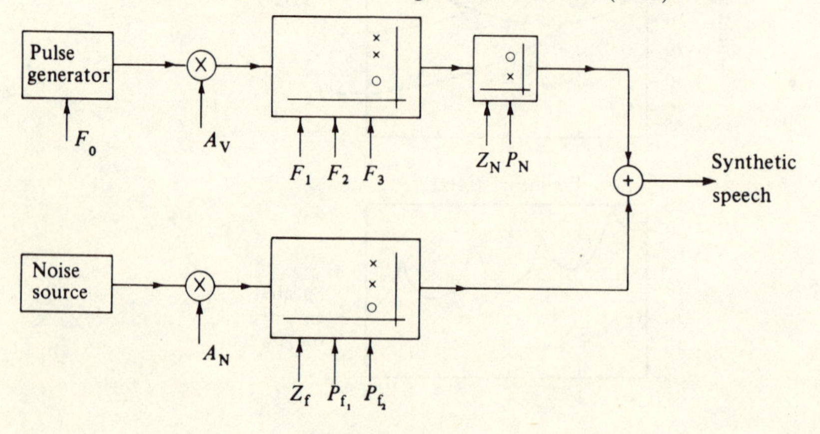

an example, consider the circuit of Fig. 4.5, due to Flanagan *et al.* (1962), which is particularly elegant. There are four resonators, only three of which are of variable frequency. Nasals are catered for by using an extra pole–zero section in the serial branch. When nasality is not required the pole and zero frequencies are set equal and therefore cancel each other. Voiceless sounds are produced by a separate branch connected in parallel, which has two poles and one zero, and is excited by random noise. There are a total of eleven control parameters, which, when varied correctly, produce high quality speech.

The advantage of the parallel formant synthesiser is that extra branches are not absolutely necessary to obtain nasals, stops and fricatives. The disadvantage is that the formant amplitudes must be specified as well as the formant frequencies. A potential problem of parallel connection of resonators is that phase cancellation at frequencies between two resonances may cause zeros in the frequency response. This effect can be explained by considering the sum of two second-order, bandpass filter functions.

$$\frac{s}{s^2 + B_1 s + \omega_1^2} + \frac{s}{s^2 + B_2 s + \omega_2^2} = \frac{s(2s^2 + sB_1 + sB_2 + \omega_1^2 + \omega_2^2)}{(s^2 + B_1 s + \omega_1^2)(s^2 + B_2 s + \omega_2^2)}$$

The numerator term will have complex conjugate roots (zeros), near the two pole frequencies ω_1 and ω_2. This will introduce an undesirable deep minimum into the frequency response between the two resonant peaks. This problem can be overcome, as Weibel (1955) has shown, by alternating the signs of the formant gains. This can eliminate the complex zeros, and Weibel gives examples of parallel resonator connections which fit vowel spectra very closely. Mathematically, the frequency response required to model a vowel spectrum can be expressed as the function $Z(s)$. If losses are neglected, and $Z(s)$ is expanded as partial fractions, then

$$Z(s) = \sum_{k=1}^{\infty} (-1)^k |A_k| \frac{s}{s^2 + \omega_k^2} = \sum_{k=1}^{\infty} (-1)^k A_k Z_k$$

where each Z_k is a simple bandpass resonator representing the formant k, and A_k is its amplitude. The $(-1)^k$ term is the alternation of signs required to avoid introducing spectral zeros.

Many formant synthesisers employing parallel connection of resonators have been reported in the literature (Cooper *et al.*, 1951; Holmes, Mattingly & Shearme, 1964; Holmes, 1973). As an example of a particularly successful design consider Fig. 4.6, which shows the schematic arrangement of the synthesiser developed by Holmes (1972). This machine uses four formant resonators connected in parallel. The signs of the formant gains are alternated to minimise the effect of zeros between formants, and

Fig. 4.6. Schematic diagram of parallel formant synthesiser. F_0 is pitch and T_0 is duration of pitch pulse; S is a voiced/unvoiced switch: A_N, A_1, A_2, A_3, A_4 are amplitudes; F_N, F_1, F_2, F_3 are formant frequencies. After Holmes (1972).

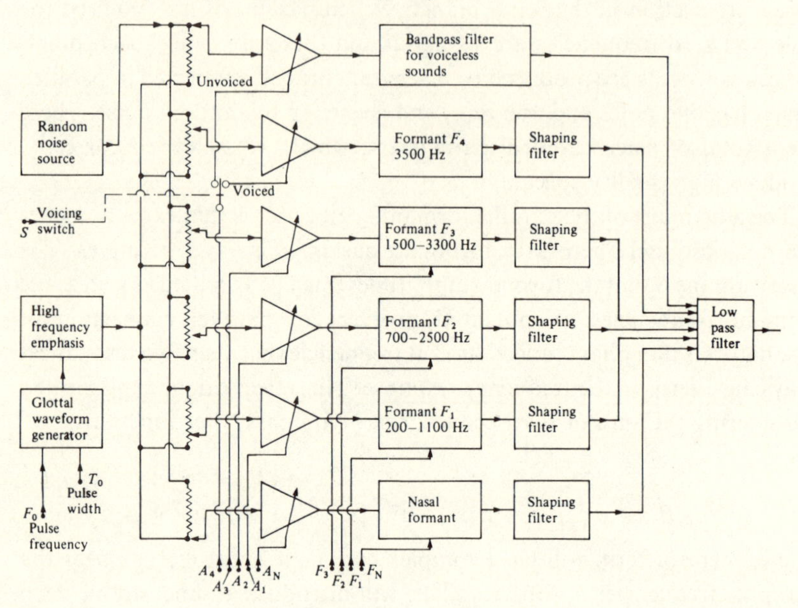

Fig. 4.7. Schematic diagram of serial–parallel formant synthesiser. F_0 is pitch, A_v, A_{vs}, A_H, A_F, A_B, and A_1 to A_6 are amplitude factors; *RGS*, *RNP*, R_1, R_2, R_3, R_4, R_5 & R_6 are resonators with variable frequency and bandwith; *RGZ*, and *RNZ* are anti-resonators with variable frequency and band width. After Klatt (1980).

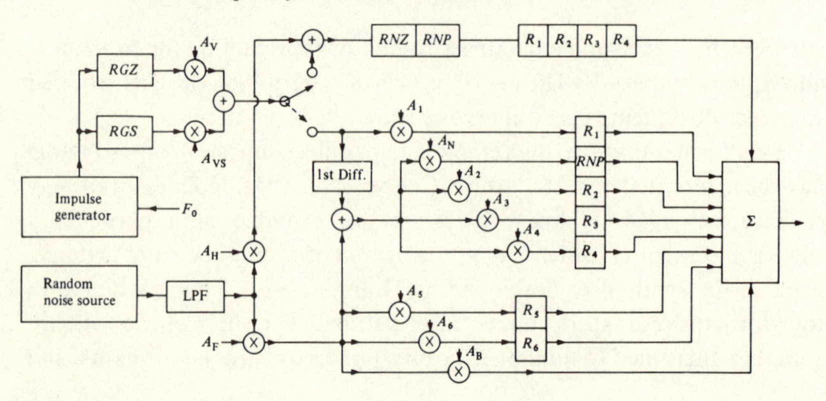

in addition each formant is followed by a spectral shaping filter. Separate parallel branches are provided for nasals and fricatives. Each formant amplitude is variable, as are the contributions from the nasal and fricative branches. The frequencies of the first three formants can be varied, and also the fundamental frequency of the excitation waveform which has a variable pulse width. The voiced/voiceless option is controlled by a binary switch. There are, altogether, twelve control signals which when varied properly are capable of giving high quality speech. It may be noted that the control requirements of the serial and parallel formant synthesisers illustrated in Figs. 4.4, 4.5 and 4.6 are not very different from each other.

A combined serial–parallel formant synthesiser developed by Klatt (1980), at MIT, is shown in Fig. 4.7. The idea here is to produce non-nasal, voiced sounds with a serial connection of resonators, and use a parallel arrangement to obtain nasals, stops and fricatives. This is a very successful strategy, even though the number of control parameters is greater than in simpler synthesisers.

4.4 Analogue circuits for formant synthesis

Circuits to implement formant synthesisers consist essentially of pulse and noise sources, exciting variable frequency filter networks. Many ingenious designs using discrete transistors and vacuum tubes have been reported in the literature, but modern integrated circuit technology has made such designs largely redundant. All the functions required for a formant synthesiser can be realised using Integrated Circuit Operational Amplifiers (Op-Amps), together with some kind of electronically-variable resistive component. The circuits given below have been found to work

Fig. 4.8. Circuit diagram of variable frequency oscillator for voiced excitation.

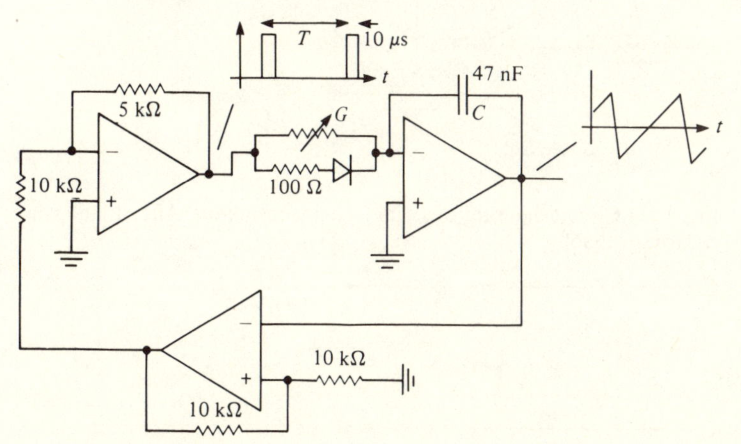

satisfactorily, but they are not claimed to be optimum in any way. All that can be said is that they can be used to implement formant synthesisers of reasonable quality at reasonable cost.

The diagram of Fig. 4.8 shows an Op-Amp circuit capable of giving variable frequency pulses suitable for voiced excitation of an analogue synthesiser. There are two outputs, one is a short pulse, the other a triangle wave. The equation for the repetition frequency is $F_0 = 10^6\,G$. Frequency control is effected by variation of the conductance G.

The circuit of Fig. 4.9 shows a random noise generator, which uses the reverse-biased base–emitter junctions of bipolar transistors as the noise source. This gives approximately white noise over a band 100–10 000 Hz.

A useful variable gain network is given in Fig. 4.10. Variation of either

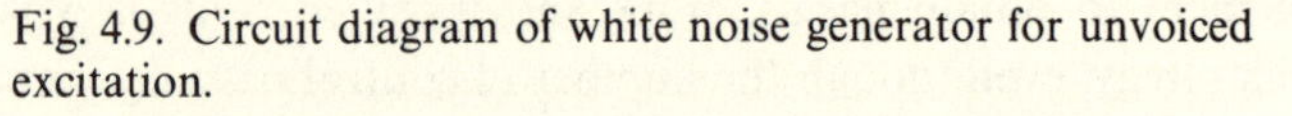

Fig. 4.9. Circuit diagram of white noise generator for unvoiced excitation.

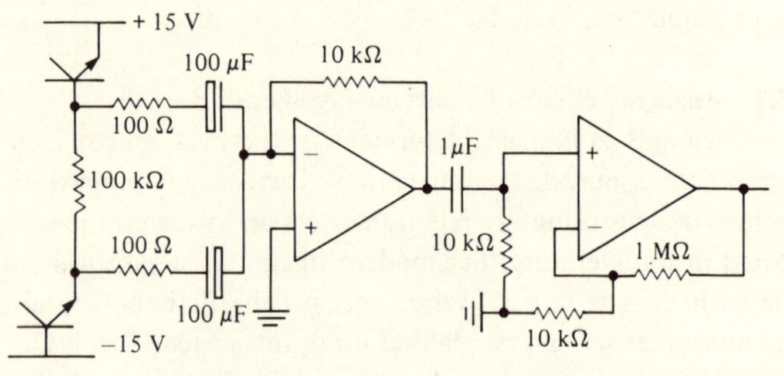

Fig. 4.10. Circuit diagram of variable gain amplifier.

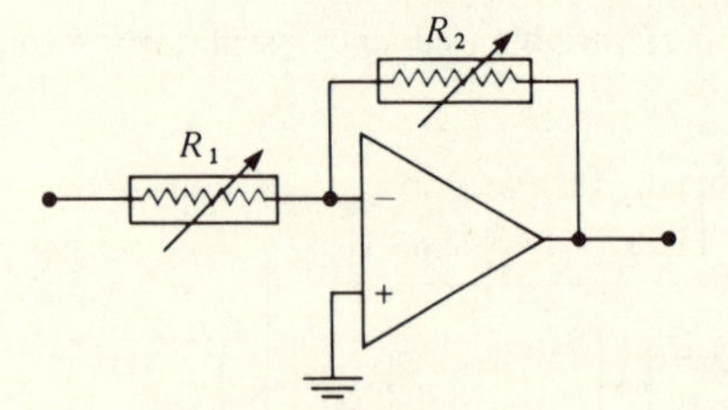

Fig. 4.11. Circuit diagram of serial formant resonator. After Flanagan & House (1956).

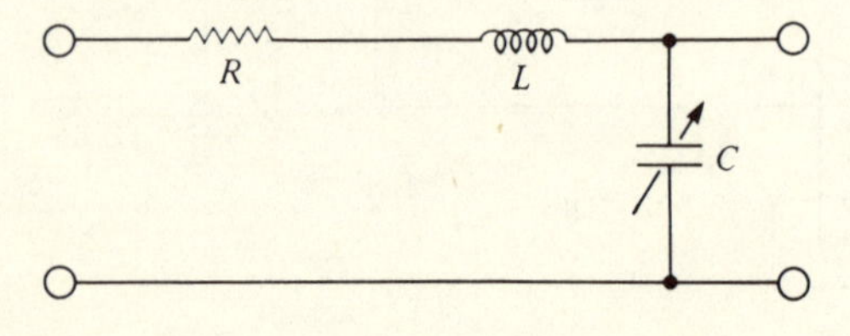

R_1 or R_2 (or both) will give variable gain according to the equation

$$\text{gain} = \frac{-R_2}{R_1}$$

A variable frequency low pass filter network is shown in Fig. 4.11. This circuit, due to Flanagan (1956), uses variable capacitance to obtain variation of filter centre frequency. The filter transfer function is given by

$$Z(s) = \frac{1/LC}{s^2 + sL/R + 1/LC}$$

The centre frequency is given by $(LC)^{-1/2}$, and the bandwidth by R/L. The variable capacitance may be realised by using variable feedback as shown in Fig. 4.12. The effective value of capacitance is $C(R_1 + R_2)/R_1$. Thus, either R_1 or R_2 can be used to control resonant frequency.

Bandpass resonators and anti-resonators, are best realised using active filter circuits. Fig. 4.13 shows a general purpose circuit using four Op-Amps, which is capable of giving bandpass, low pass and anti-resonant responses.

Fig. 4.12. Circuit diagram of variable frequency resonator.

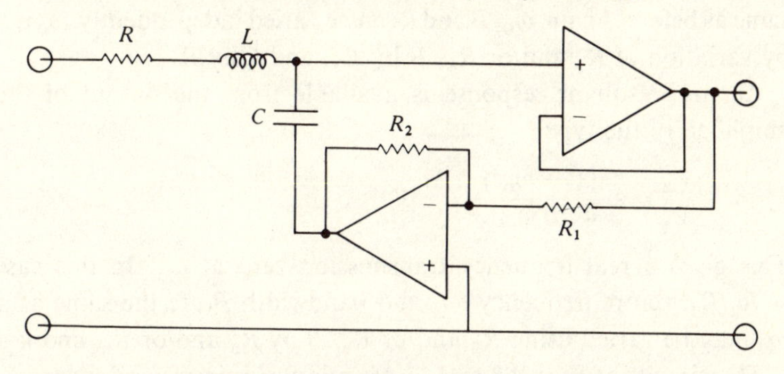

Fig. 4.13. Circuit diagram of general purpose active filter, suitable for formant synthesis. V_L is lowpass output. V_B is bandpass output. V_Z is transmission zero output.

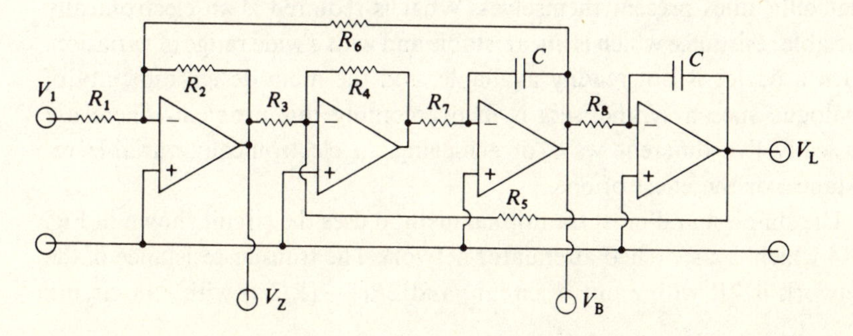

Both centre frequency and bandwidth are variable by variation of resistive components. A low pass resonance, specified by

$$\frac{V_L}{V_1} = \frac{k\omega_0^2}{s^2 + sB + \omega_0^2}$$

is available from the output of the fourth amplifier, where ω_0 is the resonant frequency, B is the bandwidth, and k is the gain factor. These parameters, given in terms of the circuit components are as follows:

$$\omega_0^2 = \frac{R_4}{R_5 R_7 R_8 C^2}, \quad B = \frac{R_2 R_4}{R_3 R_6 R_7 C}, \quad k = \frac{R_2 R_5}{R_1 R_3}$$

An examination of these expressions reveals that ω_0, B and k may be varied independently by varying resistive components. Thus ω_0 may be varied by varying R_8, k by varying R_1, and B by R_6.

A bandpass output is available from the third amplifier, giving the response

$$\frac{V_B}{V_1} = \frac{kBs}{s^2 + sB + \omega_0^2}$$

In this case $k = -R_6/R_1$, whilst centre frequency and bandwidth are the same as before. Again, ω_0, B and k can be varied independently. ω_0 is varied by variation of R_5 and/or R_8, B by R_3, and k by R_1.

An anti-resonant response is available from the output of the first amplifier, of the type

$$\frac{V_Z}{V_1} = \frac{k(s^2 + \omega_0^2)}{s^2 + sB + \omega_0^2}$$

This gives a real frequency transmission zero at ω_0. In this case $k = -R_2/R_1$, centre frequency ω_0, and bandwidth B are the same as above. ω_0 may be varied using R_5 and/or R_8, B by R_3 and/or R_6, and k by R_1.

The circuits of Figs. 4.8 to 4.13 are relatively simple and inexpensive to design and construct. It is only when it is necessary, as in the present application, to vary the circuit parameters in real time by electronic means, that difficulties present themselves. What is required is an electronically variable resistance which is linear, stable and with a wide range of variation. Such a device is not readily available, and the main design difficulty of analogue speech synthesisers is in overcoming this problem. There are basically five different ways of achieving an electronically variable resistance, or the effect of one.

The simplest and most traditional method uses the circuit shown in Fig. 4.14 which is a switched attenuator network The transfer resistance of the network is $2R$ with r out of circuit, and $2R(1 + (R/2r))$ with r in circuit.

Switching r in and out of circuit rapidly causes the transfer resistance to take up an average value depending on the fraction of time that r is in circuit. This is known as mark/space modulation and for use in a synthesiser the switching frequency must be ultrasonic, say about 100 kHz. This is not difficult to achieve and this technique has proved quite successful. The transfer conductance of the network is given by

$$G = \frac{1}{2R}\left[\frac{2r + R(1-a)}{2r + R}\right]$$

where a is the fraction of the switching cycle for which r is in circuit.

Another type of electronically variable resistance makes use of cadmium sulphide or cadmium selenide photo-resistors. These compounds conduct when exposed to light, and conductance is proportional to photon-flux over a substantial range of illumination. Thus, a light dependent resistor (LDR), in the same encapsulation as a light-emitting diode (LED), as illustrated in Fig. 4.15, makes a very effective electronically variable conductance. The main problem with such devices is their stability and response time. The decay time of this photo-electric effect is several tens of milliseconds, which severely limits the control bandwidth. Cadmium selenide has a faster response time than cadmium sulphide, and is just adequate for speech synthesis applications. These devices have a wide dynamic range, are linear and can tolerate large voltages. Indeed, they would be ideal for this application, if it were not for the fact that they are

Fig. 4.14. Switched attenuator network to give effect of variable resistance. Ratio of V_{IN}/I varies with mark/space ratio of switching waveform.

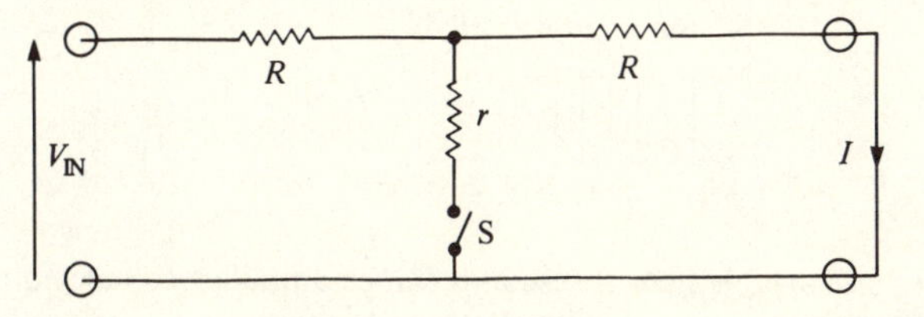

Fig. 4.15. Light Dependent Resistor (LDR), in same encapsulation as Light Emiting Diode (LED), to give current-variable resistance.

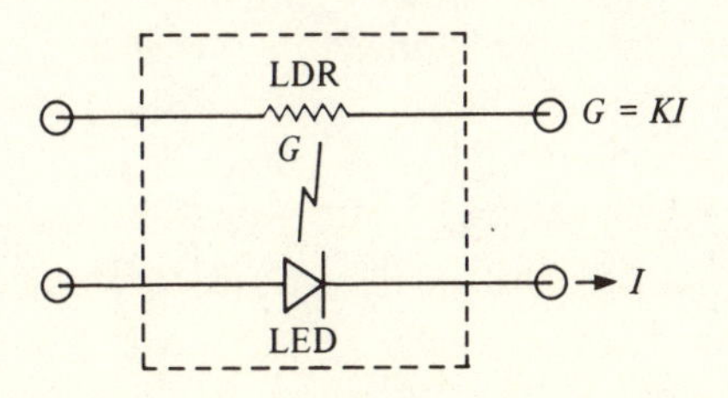

affected by temperature. However, in most applications this can be compensated, and for speech the effect may not be perceptible.

Perhaps the most satisfactory method of obtaining the effect of an electronically variable resistor is to use an integrated circuit analogue multiplier, as illustrated in Fig. 4.16. The current through the fixed resistor R_1 varies with the control voltage V_c, which is fed to the multiplier y-input. This method is stable, linear and accurate, but unfortunately expensive.

A less expensive method of achieving the same effect is to use a multiplying digital-to-analogue converter, as shown in Fig. 4.17. This scheme has the additional advantage of a digital input which facilitates control via a digital computer. The disadvantage is that it is now not

Fig. 4.16. Analogue multiplier used to create the effect of a voltage-variable resistance.

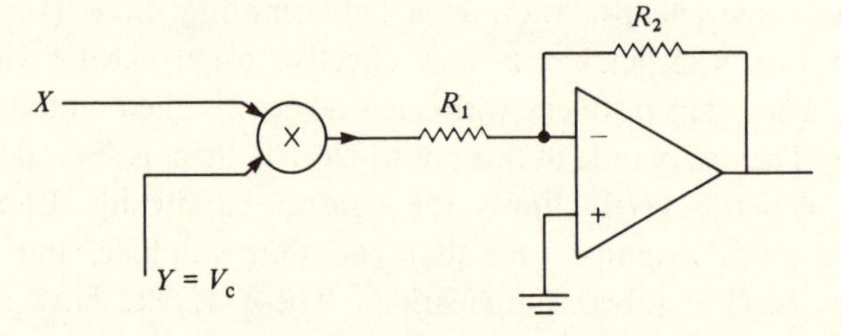

Fig. 4.17. Multiplying digital-to-analogue convertor, used to create the effect of a digitally variable resistance.

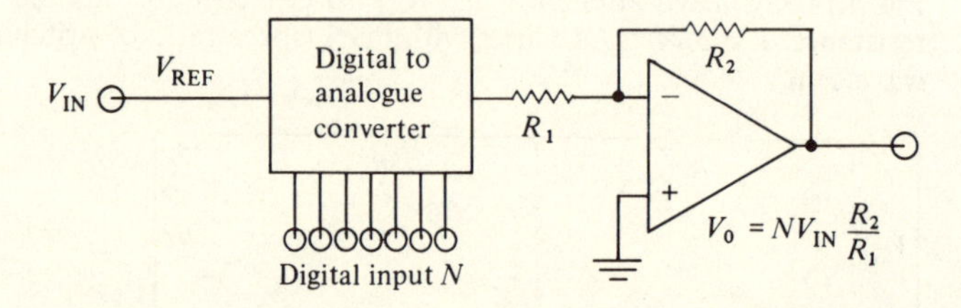

Fig. 4.18. Variable gain transconductance amplifier connected to give the effect of an electronically variable resistance.

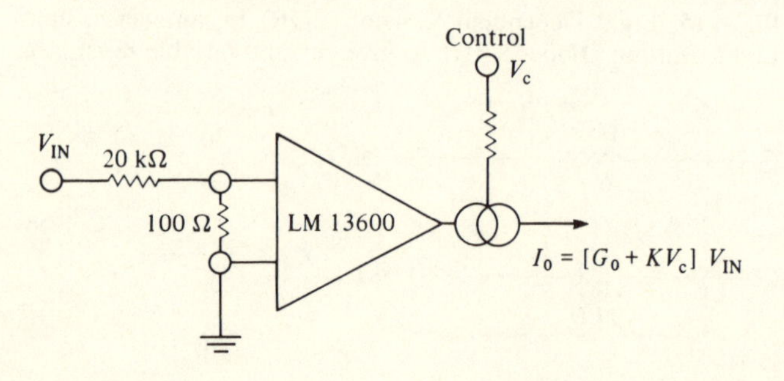

possible to introduce analogue smoothing of the control parameters – which is usually mandatory for good speech quality.

Finally, Fig. 4.18 shows a circuit which realises an electronically variable resistance by use of a variable gain transconductance amplifier. This device is particularly convenient to use, and its transconductance can be varied over several orders of magnitude. It is both stable and linear and since it is available in integrated-circuit form it is probably the most cost effective solution to the variable-filter problem.

4.5 Digital formant synthesisers

Though analogue formant synthesisers are simple and inexpensive to construct, their unreliability and potential instability make them unsuitable for professional applications. In contrast, digital synthesisers offer precision and repeatability, albeit at a somewhat higher cost.

Digital synthesisers do not necessarily have to consist of specialised digital hardware. In situations where the synthesis of speech is not required in real-time, a general purpose digital computer may be used to compute a digital filter representation of the synthesiser. The output samples of synthetic speech are calculated off-line and stored in a file. The contents of this file are then transmitted to a digital-to-analogue convertor, at an appropriate rate (typically 10 kHz), to give speech output in real-time. However, even with a fairly fast digital computer, the time taken to prepare the output file is many times real-time. This method of synthesis is therefore most useful in research applications, since the structure of the synthesiser, defined entirely by program (software), may be changed very quickly. Several such programs are given in the literature (Flanagan *et al.*, 1962; Rabiner, 1968*a*; Holmes, 1973; Klatt, 1980).

On the other hand, hardware digital synthesisers (Rabiner & Schafer, 1976), are designed to produce speech in real-time. Usually, they are fixed in structure and once built are usually difficult to change, though some programmable machines have been built (Linggard & Marlow, 1979). However, the theoretical background to the design of digital synthesisers is common to both software and hardware realisations, thus the design techniques described below may be interpreted as either computer algorithms or digital circuit diagrams.

In a digital formant synthesiser, the filtering is accomplished by sampled data filters. After the sampling of a signal, it is usual to quantise (digitise) the sampled data, and consequently, such filters are commonly referred to as 'digital filters'. The theory of sampled data (digital) filtering utilises the z-transform domain, rather than the Laplace domain of continuous data (analogue) filters. Filter functions are thus functions of the variable z, rather than s, the Laplace operator. The relationship between z and s is given by

$$z = e^{sT}$$

where T is the sampling period. Functions of z can thus be expressed in terms of the frequency variable ω, and frequency responses may be plotted in the usual way. The sampling period T may be any convenient value under the Nyquist restriction that the sampling frequency $1/T$ must be at least twice the highest frequency of interest. For speech this is about 5 kHz, thus the sampling frequency is typically 10 kHz, giving $T = 100\,\mu s$.

In a digital filter the storage elements equivalent to the inductance and capacitance of an analogue filter, are pure time delays $z^{-1} = e-^{j\omega T}$. Fig. 4.19 shows a simple digital filter using one delay in a feedback loop. The transfer function of this configuration is given by

$$H(z) = \frac{1}{1 - az^{-1}}$$

which can also be expressed as a function of ω, as

$$\frac{1}{1 - ae^{-j\omega T}} = \frac{1}{1 - a(\cos \omega T - j \sin \omega T)}$$

Fig. 4.19. Simple lowpass digital filter, using a single delay element.

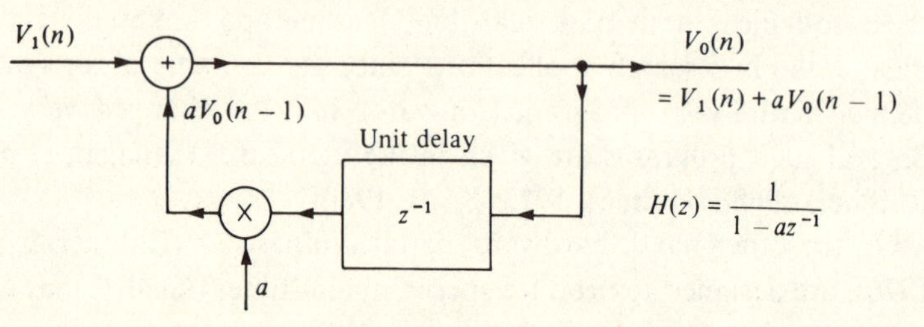

Fig. 4.20. Frequency response of the simple lowpass digital filter, shown repeating at intervals of half the sampling frequency.

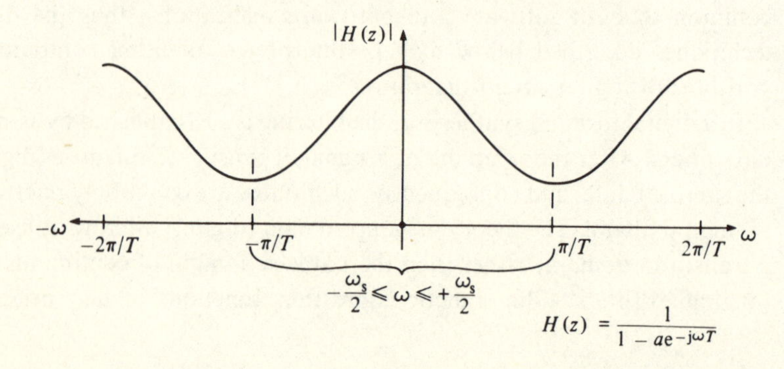

The magnitude of this function is

$$[(1 - a\cos\omega T)^2 + (a\sin\omega T)^2]^{-1/2}$$

Over the range $-\pi < \omega T < +\pi$, this function has the magnitude response shown in Fig. 4.20. This shape repeats every integer multiple of 2π.

The periodic nature of the frequency response of digital filters is due to the identity $z^{-1} = e^{-j\omega T} = \cos\omega T - j\sin\omega T$, so that functions of z repeat every time ωT cycles through integer multiples of 2π. Since $T = 2\pi/\omega_s$, the frequency response of functions of z will be unique within the range $-\omega_s/2 < \omega < +\omega_s/2$, where ω_s is the radian sampling frequency. That is, the frequency response is unique for ω within half the sampling frequency, a condition which is already required of the signal by the Nyquist criterion. Thus, within the frequency range of interest, the periodic nature of $H(z)$ does not cause problems.

In the expressions characterising digital filters, it is convenient to normalise frequency ω by scaling to the sampling frequency ω_s. Thus, normalised frequency $\theta = 2\pi\omega/\omega_s$, and since ω will lie within the range $\pm\omega_s/2$, θ will lie within $\pm\pi$. As ω goes from 0 to $\omega_s/2$, θ goes from 0 to π; and as ω goes through integer multiples of $\omega_s/2$, θ cycles through integer multiples of π. Under the Nyquist restriction $-\omega_s/2 < \omega < +\omega_s/2$, θ will only extend to $\pm\pi$, that is, once around the unit circle.

The frequency response shown in Fig. 4.20, in normalised form is $1/(1 - ae^{-j\theta})$ and in the range $0 < \theta < \pi$ the response is low pass.

The relationship between s and z means that there is a mapping between the s-plane and the z-plane. This is shown in Fig. 4.21. The origin of the s-plane, $s = 0$, maps to $z = +1$. The $j\omega$ axis maps to the unit circle, and the whole left half of the s-plane maps to within the unit circle. Since the poles of s-plane functions which lie in the left half of the s-plane are stable, poles

Fig. 4.21. s-plane to z-plane mapping. The left-half plane maps to within the unit circle of the z-plane.

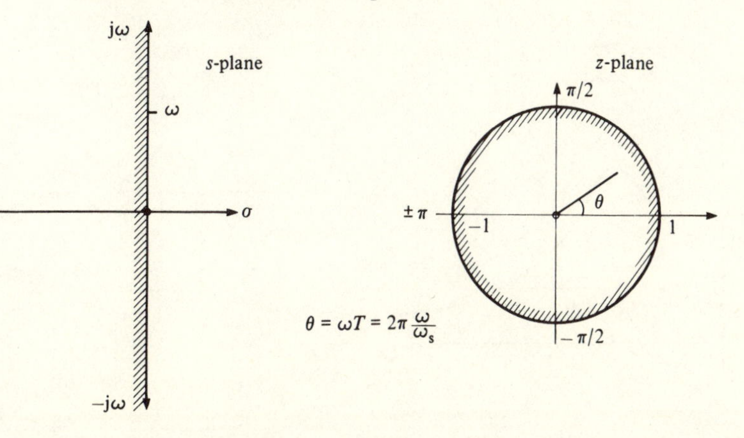

of z-plane functions which lie within the unit circle will also be stable. The pole of the function

$$H(z) = \frac{1}{1 - az^{-1}}$$

is at $z = +a$, which lies within the unit circle if $a < 1$. Thus, this function of z will be stable only if a is less than unity.

Complex conjugate poles in the z-plane may result from using two delays in a feedback loop, as shown in Fig. 4.22. Here the transfer function is

$$H(z) = \frac{k}{1 + b_1 z^{-1} + b_2 z^{-2}}$$

If the complex conjugate poles are at $z = Re^{-j\phi}$, and $z = Re^{+j\phi}$, the denominator may be written

$$(1 - z^{-1}Re^{-j\phi})(1 - z^{-1}Re^{+j\phi})$$
$$= 1 - 2Rz^{-1}\cos\phi + R^2 z^{-2}$$

Thus $b_1 = -2R\cos\phi$, and $b_2 = R^2$, or given b_1 and b_2, $R = (b_2)^{1/2}$, and $\cos\phi = -b_1/2(b_2)^{1/2}$.

The magnitude response of this function is shown in Fig. 4.22. The factor k has been chosen to make $|H(z)| = 1$, at zero frequency, that is $k = 1 +$

Fig. 4.22. Two-delay digital filter giving complex conjugate poles in the z-plane. Frequency response is lowpass resonant.

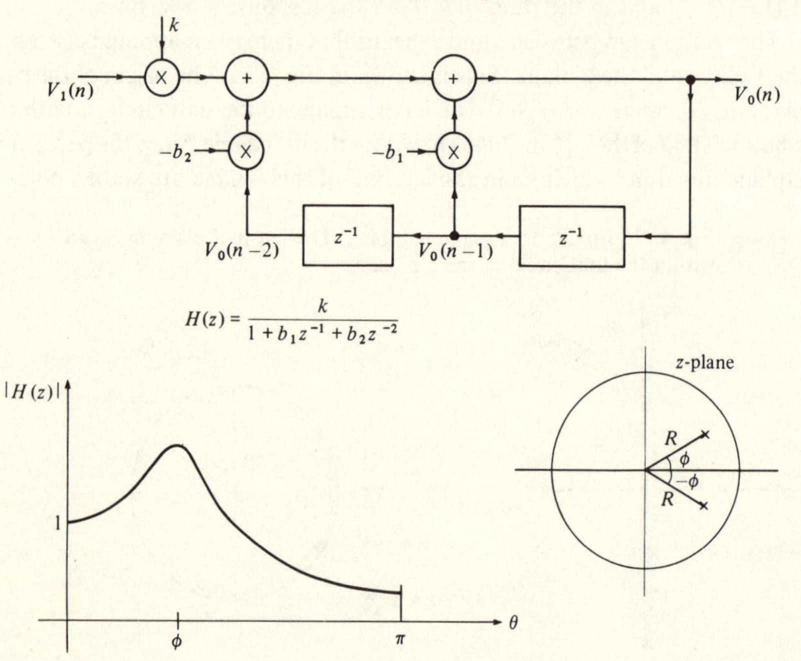

$b_1 + b_2$. The function is low pass with a resonant peak near $\theta = \phi$. The bandwidth of the resonance is governed by the pole radius R, and as $R \rightarrow 1$ the bandwidth approaches zero. The actual relationship is $R = e^{-\alpha}$, where α is defined as the normalised bandwidth in radians.

The digital filter diagram of Fig. 4.22 may be interpreted as either a computer algorithm, or a schematic representation of a digital circuit. The equation for $H(z)$ can be translated to a computer algorithm as follows:

$$\frac{v_0(n)}{v_1(n)} \rightarrow \frac{V_0(z)}{V_1(z)} = \frac{k}{1 + b_1 z^{-1} + b_2 z^{-2}}$$

where $v_1(n)$ and $v_0(n)$ are the input and output quantities of the filter at the discrete time interval n. This may be expanded as

$$kV_1(z) = V_0(z) + b_1 z^{-1} V_0(z) + b_0 z^{-2} V_0(z)$$

Since $z^{-1}V_0(z) = v_0(n-1)$, and $z^{-2}V_0(z) = v_0(n-2)$, the equation may be written

$$v_0(n) = kv_1(n) - b_1 v_0(n-1) - b_2 v_0(n-2)$$

That is, the present output $v_0(n)$ can be represented as a weighted sum of present input and past outputs. This expression must be calculated for every time interval n, with the past outputs being updated at each step. This can be done quite simply on a digital computer, with four or five lines of high level language. Alternatively, a modern signal processing chip could be configured to perform this calculation in a few microseconds.

Digital resonators of this type are very useful in the design of serial-

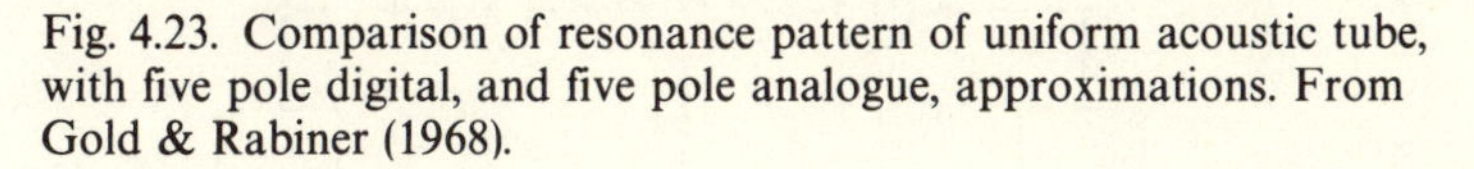

Fig. 4.23. Comparison of resonance pattern of uniform acoustic tube, with five pole digital, and five pole analogue, approximations. From Gold & Rabiner (1968).

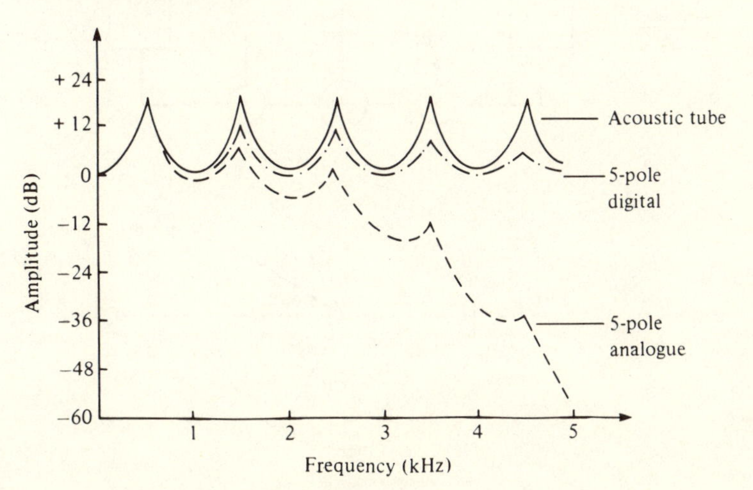

formant synthesisers, since they are low pass and may have unity d.c. gain. There is an additional advantage, in this application, however, in that no higher pole correction is necessary. The reason is that the periodic nature of the digital filter transfer function effectively puts in an infinity of higher poles, and extra compensating poles are not required. Gold & Rabiner (1968) and Rabiner (1968*a*), have calculated the precise extent of this correction, and Fig. 4.23 shows a comparison of five-formant vowel spectra from an acoustic tube model, from an analogue filter, and from digital resonators. It can be seen that the digital realisation is very close to the acoustic tube ideal.

To cater for the production of nasals, fricatives and stops in a serial synthesiser, it is necessary to use functions which give both poles and zeros in the *z*-plane. Fig. 4.24 shows a digital filter which realises a complex pair of poles and zeros. The transfer function is

$$H(z)=\frac{k_1}{k_2}\frac{1+a_1z^{-1}+a_2z^{-2}}{1+b_1z^{-1}+b_2z^{-2}}$$

Fig. 4.24. Complex pole–zero digital filter, using four delay elements.

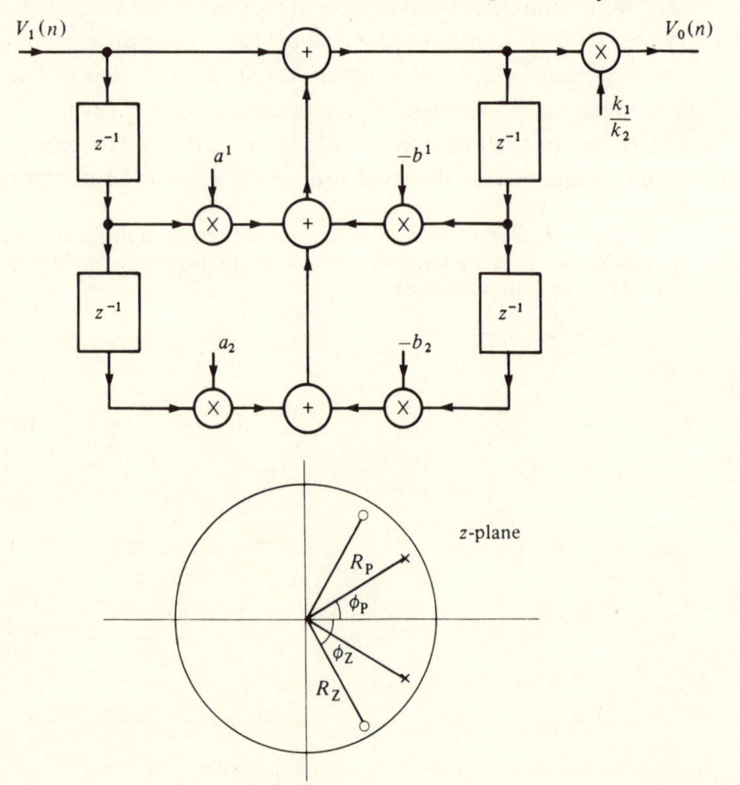

As before the pole parameters b_1 and b_2 are given by

$$b_1 = -2R_p \cos\phi_p \quad \text{and} \quad b_2 = R_p^{\,2}$$

The zero parameters a_1 and a_2 are given by

$$a_1 = -2R_z \cos\phi_z \quad \text{and} \quad a_2 = R_z^{\,2}$$

where ϕ_p and ϕ_z are the normalised pole/zero frequencies, and R_p and R_z are their radii. The constants k_1 and k_2 are introduced to make the function equal to unity at d.c. That is,

$$k_1 = 1 + b_1 + b_2, \quad \text{and} \quad k_2 = 1 + a_1 + a_2$$

The complete schematic diagram of a typical, digital serial-formant synthesiser is shown in Fig. 4.25. This is the Bell Laboratories machine, described in detail by Rabiner, Schafer & Coker (1971).

Digital, parallel-formant synthesisers consist of bandpass resonators connected in parallel. In the sampled data case, band-pass implies a function having zeros at $\omega = 0$, and $\omega = \omega_s/2$, that is, for normalised frequency θ, the function has zeros at $\theta = 0$, and $\theta = \pi$. A function of z which has this property is

$$H(z) = \frac{a_0(1 - z^{-1})(1 + z^{-1})}{1 + b_1 z^{-1} + b_2 z^{-2}} = \frac{a_0(1 - z^{-2})}{1 + b_1 z^{-1} + b_2 z^{-2}}$$

The $(1 - z^{-1})$ factor gives the zero at $\theta = 0$, and the $(1 + z^{-1})$ factor gives zeros at $\theta = \pi$. The denominator factor provides a resonant peak at $\theta = \phi_p$, with a normalised bandwidth of $\Delta\phi$. The coefficient a_0 is used to make this peak value equal to unity. The significance of these parameters is shown in

Fig. 4.25. Schematic diagram of typical digital formant synthesiser. After Rabiner, Schafer & Coker (1971).

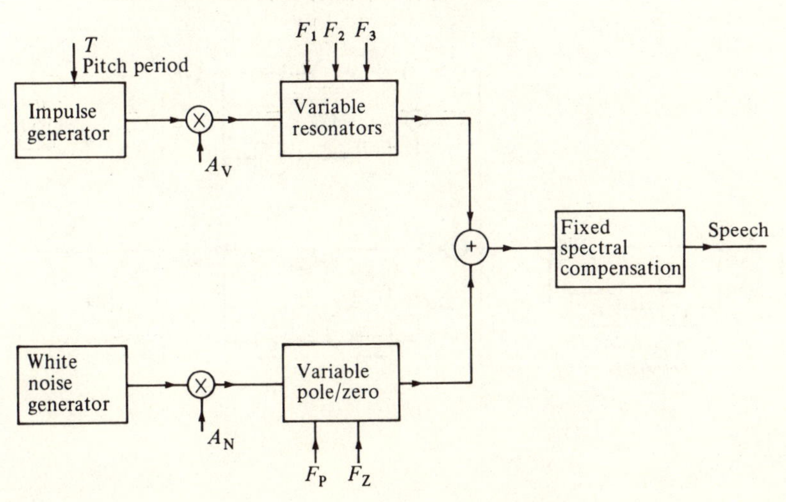

Fig. 4.26. The design problem is to specify the filter coefficients a_0, b_1 and b_2 in terms of the required performance parameters ϕ_p and $\Delta\phi$. It can be shown that

$$a_0 = \frac{t}{1+t}, \quad b_1 = \frac{-2\cos\phi_p}{1+t} \quad \text{and} \quad b_2 = \frac{1-t}{1+t}$$

where $t = \tan\frac{1}{2}\Delta\phi$.

For most applications of parallel formant synthesis, the bandwidths of the formants can be fixed, and only the formant frequencies and their

Fig. 4.26. Bandpass digital resonator for parallel digital formant synthesiser.

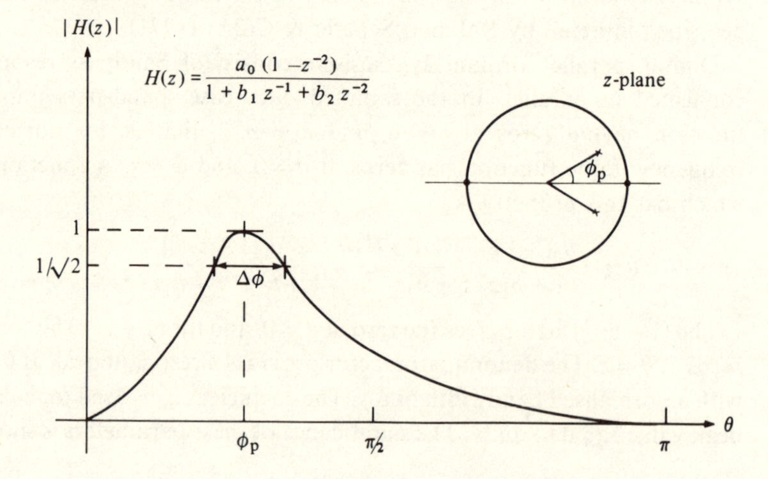

Fig. 4.27. Arrangement of digital, parallel-formant synthesiser.

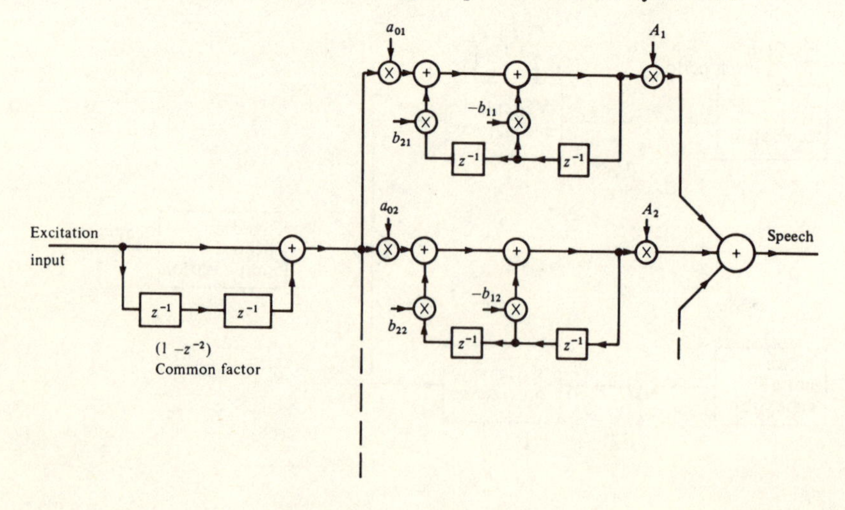

amplitudes need be varied. Thus, both a_0 and b_2 may be fixed, leaving b_1 as the formant frequency control parameter. This type of bandpass function is particularly easy to incorporate into a computer algorithm or implement in digital hardware, and Fig. 4.27 shows a typical arrangement. Each formant filter has only two control parameters, the amplitude factor k, and the formant frequency coefficient b_1. The factor $(1 - z^{-2})$ is common to all the filter functions, so that it need only be computed once and its output fed to the resonators. Furthermore, if the formant bandwidths are all set equal, then the a_0 and b_2 coefficients will be the same for each filter.

Digital, parallel-formant synthesisers have been described in the literature (Holmes, 1973), and a complete computer program has been given by Klatt (1980).

4.6 Excitation of terminal analogue synthesisers

The simplest way of exciting the resonant modes of a system is to stimulate it with a short pulse. This has the advantage that if the pulse is short enough then all the resonant modes will be excited with equal energy. The magnitude spectrum of a single pulse of duration ΔT is shown in Fig. 4.28(*a*), and is given by the expression

$$\left| A\Delta T \frac{\sin(\omega\Delta T/2)}{\omega\Delta T/2} \right|$$

Fig. 4.28. (*a*) Spectrum of a single 'short' pulse. (*b*) Spectrum of a periodic train of 'short' pulses.

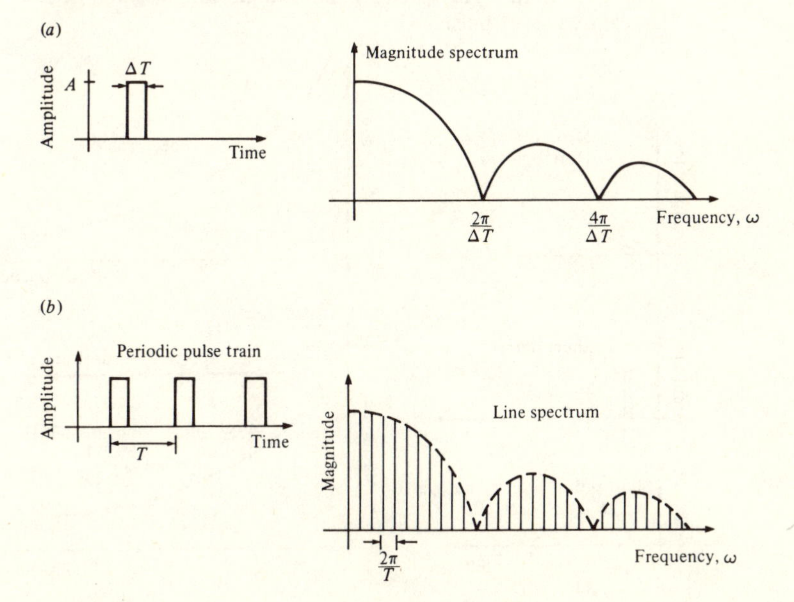

where A is the amplitude of the pulse. This has the classic $\sin(x)/x$ shape, with real frequency zeros at $\omega = 2\pi N/\Delta T$, $N = 1, 2, 3$ etc. Thus, to provide equal excitation to all resonant modes, the first zero, at $2\pi/\Delta T$, must be well above the highest resonant frequency of the system. For application in speech synthesis, with resonant modes of interest up to 5 kHz, placing the first zero at 50 kHz gives a reasonably flat spectrum over the required frequency band. A 'short' pulse in this context, therefore, would have a duration of about 20 μs.

If such a pulse is to be used to give periodic excitation for voiced speech, it must, of course, be repeated at pitch frequency. The effect of this repetition is to cause the spectrum to become harmonic, with harmonic components spaced at pitch intervals. If there are an infinite number of repetitions, then the harmonics will become 'lines', as illustrated in Fig. 4.28(*b*). The $\sin(x)/x$ shape then becomes the 'envelope' of this harmonic line spectrum.

It is usual to assume that, if the spectrum of the synthetic speech can be made to match that of real speech, then there will be no perceptible difference. This is not quite true, since at low frequencies the ear is phase sensitive (Holmes, 1973). Thus, though it is possible to create voiced excitation spectra which have the correct amplitude characteristics, the synthesised speech will only sound natural if the time waveform of the excitation pulses is similar to that of natural speech. For example, the glottal pulse wave shape shown in Fig. 4.29 has an amplitude spectrum

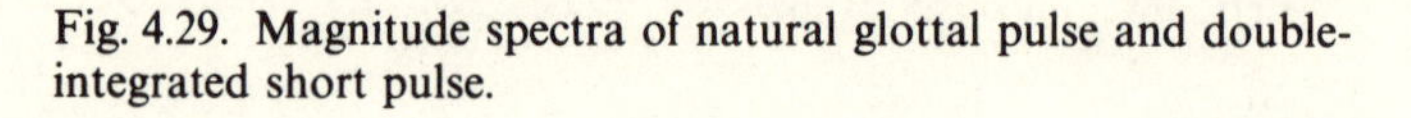

Fig. 4.29. Magnitude spectra of natural glottal pulse and double-integrated short pulse.

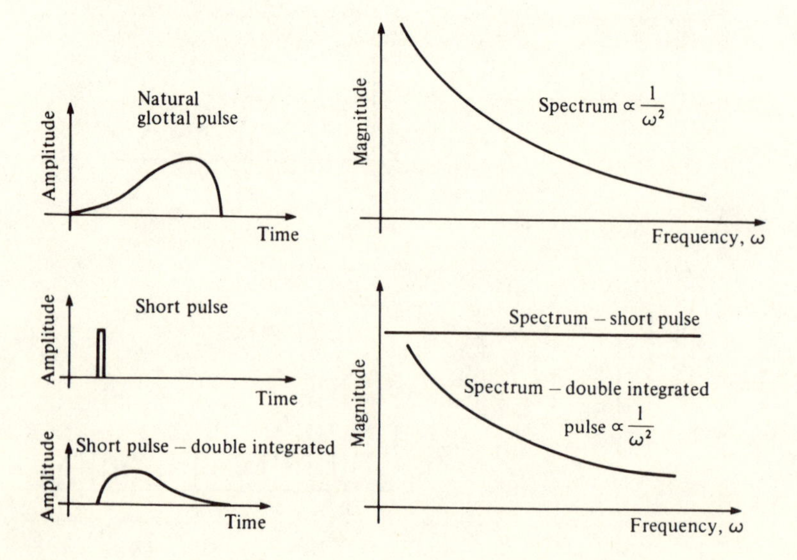

falling off at a rate proportional to $1/\omega^2$. To create a similar spectrum, it is only necessary to double integrate the short pulse of Fig. 4.28. Since double integration of the short pulse is impractical, a good approximation is to pass it through an over-damped, two-pole, low pass filter; the result of this is shown in Fig. 4.29. Though this wave shape has the right magnitude spectrum, its phase characteristic is not correct. The pulse is, in fact, approximately the time reverse of the natural glottal pulse shape. Thus, though the magnitude of its spectrum is identical to the natural one, the phase is reversed. It is possible, by careful listening, to distinguish between the two types of excitation, though for most purposes the time reversed pulse is quite satisfactory. It does prove, however, that the ear is indeed phase sensitive at low frequencies.

The short pulse is such a convenient waveform to generate that it has been employed in many synthesisers, especially those using analogue circuitry. The usual technique is to use a low pass filter to approximate one integration and allow the second integration to be cancelled by the lip-radiation factor which is approximately a differentiation.

A more precise approach is to use a triangular waveform to approximate the wave shape of the natural glottal pulse. The optimum shape for such pulses has been investigated by Dunn, Flanagan & Gestrin (1962), who have shown that a slightly asymmetrical triangular pulse is necessary to avoid spectral zeros. This can be appreciated by considering a triangular pulse to be an integration of two pulses of opposite polarity, as shown in Fig. 4.30. If the rise and fall times of the triangular wave are equal, $\Delta T_1 = \Delta T_2 = \Delta T$, then its spectrum is given by

$$4A \frac{\mathrm{e}^{-\mathrm{j}\omega} \sin(\omega \Delta T/2)}{-\mathrm{j}\omega^2}$$

This has the correct average fall-off, proportional to $1/\omega^2$; however, it has

Fig. 4.30. Spectrum of asymmetrical triangular pulse.

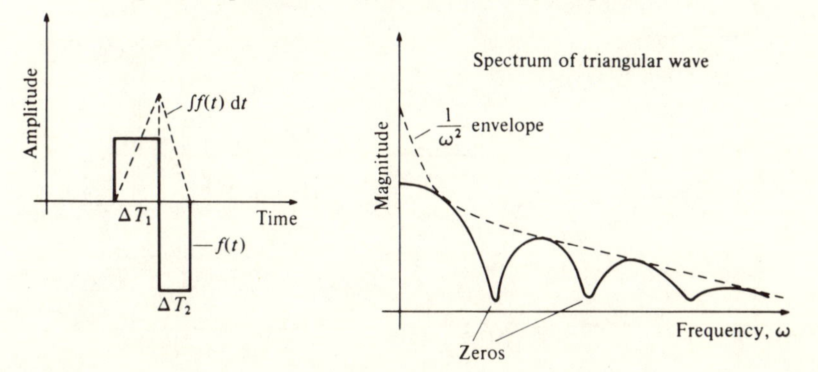

zeros at $\omega = 2\pi N/\Delta T$, $N = 1, 2, \ldots$. In this case the pulse cannot be 'short', since the duration of the triangle must be about 2 ms. Thus, there will be about five zeros in the frequency range of interest up to 5 kHz. Dunn, Flanagan & Gestrin (1962) have shown that if the triangle wave is made asymmetrical, the zeros move away from the jω axis in the s-plane. and cause only dips in the spectrum rather than complete nulls. The optimum ratio of rise and fall times is 11 to 12.

An ideal method of exciting speech synthesisers would be to use exact replicas of the natural glottal pulse. However, this is difficult to do, especially with analogue circuits. In an interesting experiment, Rosenberg (1971) replaced the original glottal wave of real utterances with various stylised versions, including triangles. Listening tests indicated that the most preferred artificial wave shape was one made up from cosine segments, as shown in Fig. 4.31, which is a good approximation to the natural shape. The amplitude spectrum of this wave shape is also a good approximation to the required $1/\omega^2$ frequency fall-off.

In generating glottal pulses synthetically, it is most convenient to create a basic shape and to repeat this at pitch frequency. The alternative is to create a shape that shrinks and expands in proportion to pitch period. There is a subtle difference in the spectra of these two possibilities. In the

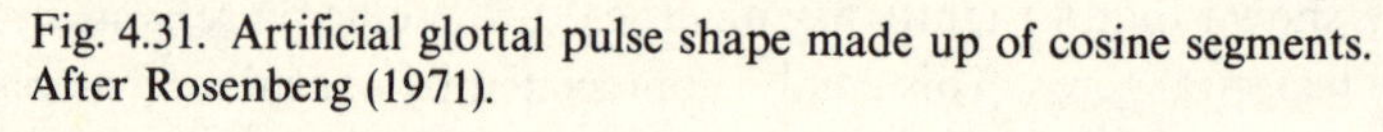
Fig. 4.31. Artificial glottal pulse shape made up of cosine segments. After Rosenberg (1971).

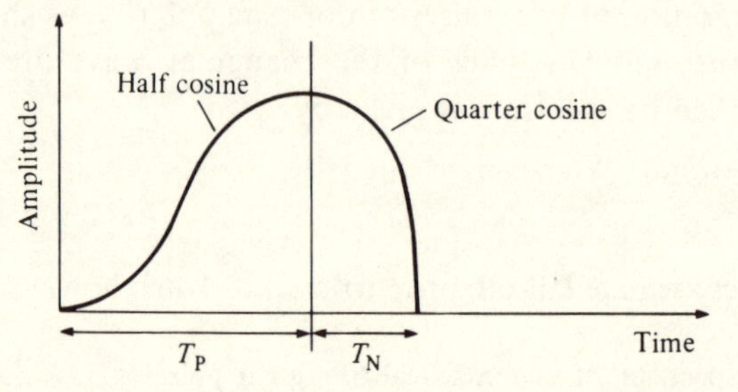

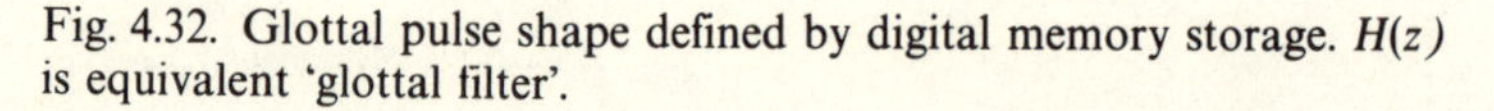
Fig. 4.32. Glottal pulse shape defined by digital memory storage. $H(z)$ is equivalent 'glottal filter'.

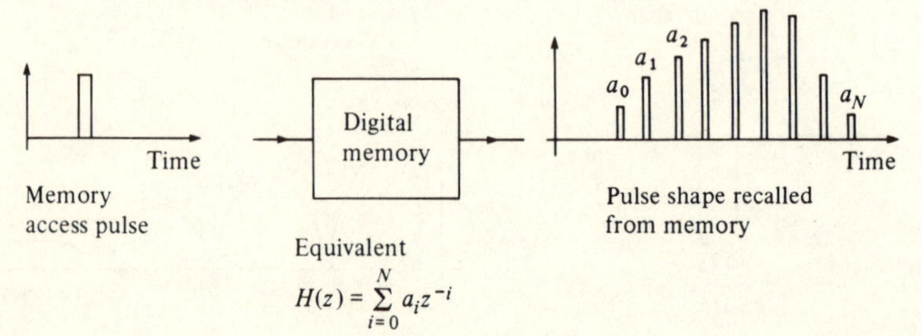

first, the fixed shape gives static zeros in the s-plane, the second method gives moving zeros. In real speech the glottal wave shape shrinks and expands with pitch period, though not in exact proportion. In synthetic speech it appears not to matter too much which method is used, so long as there are no fixed, real-frequency zeros in the glottal spectrum.

In digital synthesis, the problem of creating a complex wave shape is less troublesome, since the details of the shape can be stored in digital memory and accessed for each pitch pulse. Fig. 4.32 shows this simple arrangement. It can be shown that accessing the memory is equivalent to exciting a finite impulse response (FIR) digital filter with an impulse. If the wave shape samples are $a_0, a_1, a_2, \ldots, a_N$, then the unit pulse response of the equivalent FIR filter is simply $a_0 + a_1 z^{-1} + a_2 z^{-2} + \cdots + a_N z^{-N}$. This FIR 'glottal filter' is thus $G(z) = \Sigma_{i=0}^{N} a_i z^{-i}$. The zeros of this function are obtained as the roots of $G(z)$. In fact they are directly controllable if $G(z)$ is written as $G(z) = \Pi_{i=1}^{N} a_0(1 - b_i z^{-1})$, and the a coefficients are calculated by expanding this expression. Since a typical glottal pulse has a duration of about 5 ms and a digital synthesiser will have a sampling rate of about 10 kHz, it is feasible to represent the glottal pulse shape by about 50 samples. This is quite a detailed representation and adequate for most synthetic speech.

Turbulent excitation of speech synthesisers is usually accomplished by using some kind of random noise. In synthesisers using analogue circuitry, it is convenient to make use of the noise generated across a semi-conductor diode in its breakdown mode, as illustrated in Fig. 4.9. This is both simple and direct; however, in digital synthesis more formal methods have to be employed. Short pulses, as defined above, can be used in two different ways. Either the pulses can be generated with equal amplitudes at random intervals, or they can be repeated at equal intervals (usually the sampling

Fig. 4.33. Pseudo-random binary sequence generated by shift register and exclusive-OR gates. After Rabiner, Schafer & Coker (1971).

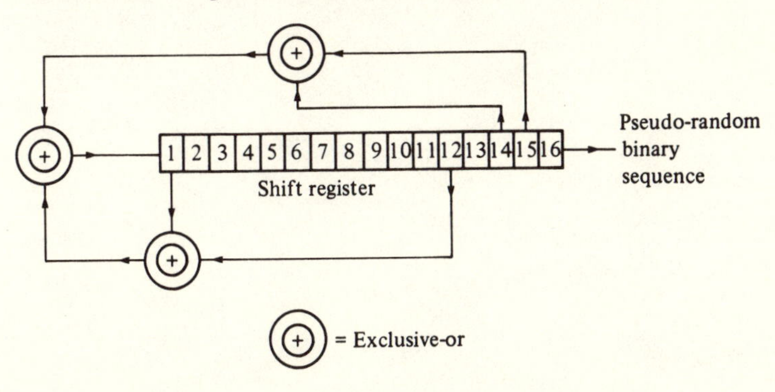

period) with random amplitudes. The effect is essentially the same in both cases; a wideband, flat spectrum with random phase is created.

A standard method useful in hardware synthesisers is to generate a pseudo-random binary sequence, using shift registers. Fig. 4.33 shows one such method used by Rabiner, Schafer & Coker (1971). It utilises a maximum length, sixteen-bit code. The output sequence of ones and zeros is interpreted by the hardware as plus and minus one, respectively, so that the mean value becomes zero.

A second method, useful in digital computer simulations of synthesisers, uses a regular sequence of samples with pseudo-random amplitudes. The easiest way of producing random amplitudes is to add numbers successively into an accumulator, permitting twos-complement overflow. That is, when a number becomes too large for the number system, it goes through the 'top' of the accumulator and back in at the 'bottom', thus becoming negative. This is essentially modulo arithmetic, and if the set of starting numbers is carefully chosen, the process can yield long sequences of pseudo-random numbers before repeating. Rabiner & Gold (1975) have used as many as 50 starting numbers, though as few as two (Linggard & Marlow, 1979) appears to give reasonable results. To be effective, a new number must be generated every sampling interval. The output then appears to be sampled random noise.

5 *Synthesis in the time domain*

Modelling the time waveform of the speech signal

'Because the transcendental functions for the vocal tract are meromorphic, and because their numerator and denominator components are generally integrable functions (i.e. analytic for all finite values of the complex variable), it is possible to approximate the transmission by rational functions.'
J.L. Flanagan, 1972

5.1 Introduction

A signal model of speech can be created in either the time domain or the frequency domain. That is, the modelling process can attempt to model the spectrum of the speech as discussed in chapter 4, or it can model the speech waveform itself. Traditionally, speech synthesis techniques have been applied to the modelling of the spectrum of speech, and only one early attempt to create speech-like waveforms directly is reported in the literature.

This is the synthesis procedure of Lawrence (1953) which uses a modulation technique to synthesise both voiced and unvoiced waveforms. The arrangement of the Lawrence synthesiser is shown in Fig. 5.1. For voiced speech, an envelope function is generated by exciting a low pass filter with short pulses, periodic at pitch frequency. This gives a sequence of exponentially decaying pulses which are then used to modulate a 10 kHz sine wave. Individual formant waveforms are created by heterodyning this modulated 10 kHz waveform with a sine wave at a frequency (10 kHz + F); where F is the frequency of the required formant. The heterodyning process translates the 10 kHz waveform down to frequency F, without altering the exponentially decaying envelope. This is done for each of the three formants using separate oscillators to heterodyne with the 10 kHz waveform. Thus all the formants have the same decay rate and hence the same bandwidths.

For unvoiced speech, random noise is low pass filtered to 100 Hz and used to modulate the 10 kHz sine wave. This creates a narrow band of noise about 10 kHz, which is translated by heterodyning with the outputs of the three sine wave oscillators. This gives three fricative formants. Lawrence synthesised various simple phrases using hand-made parameters painted on glass plates. Digitising these parameters he demonstrated synthetic speech at a control data rate of 550 bits per second.

Fig. 5.1. Lawrence's speech synthesiser. All waveforms are derived from the 10 kHz oscillator by heterodyning and filtering. After Lawrence (1953).

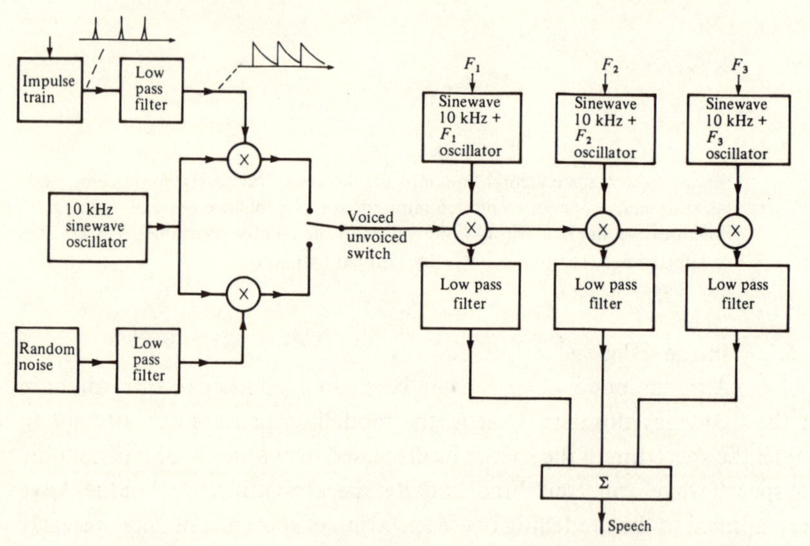

Fig. 5.2. The ICL synthesiser. Voiced speech is generated from sinewaves stored in ROM. Unvoiced speech is derived from a pseudo-random binary sequence, convolved with a suitable impulse response. After Underwood & Martin (1976).

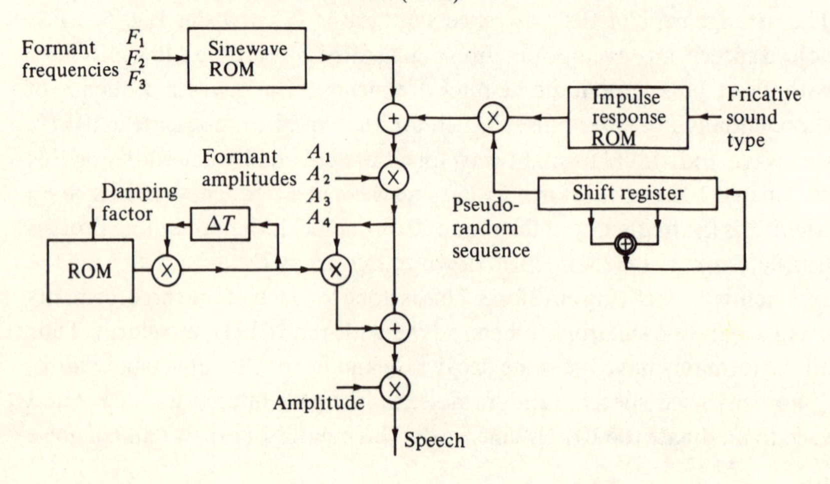

A modern version of Lawrence's machine, using digital techniques, has been developed by Underwood & Martin (1976) and is illustrated in Fig. 5.2. In this machine, the decaying envelope function for voiced sounds is created by iteratively multiplying by a decay factor and resetting at pitch frequency. The decay factors are obtained from a read-only memory (ROM) look-up table. The formant frequency sine waves are also obtained from ROM tables and added together and scaled before being multiplied by the pitch envelope. For unvoiced sounds, five different impulse responses are stored in memory and accessed as required. Each impulse response represents one of the five fricative sounds of English. That is, the Fourier transform of each impulse response is an approximation to the spectrum of the corresponding fricative. The actual sounds are generated by convolving a stored impulse response with a pseudo-random sequence. Since this sequence is binary, the convolution consists of simple gating of the impulse response samples. The hardware required for this synthesiser is very simple indeed and may operate at very high speed. The machine was, in fact, designed to be multiplexed by a factor of 32.

Other direct waveform synthesis methods have been reported which use recorded segments of real speech in a phoneme synthesis procedure, but details are not available, and their success is doubtful. In general, the modelling of the speech time waveform is fairly easy for a low level of speech quality, and simple circuits suffice to achieve this; but for good quality, highly intelligible, synthetic speech, only Linear Predictive Coding (LPC) offers an acceptable solution. The speech synthesis model which results from the LPC method is an all-pole digital filter – similar to the serial formant synthesiser of chapter 4. But the essence of LPC is that it attempts to recreate the time waveform of original speech rather than its spectrum. Because LPC approximates the time waveform of the original speech very closely, the spectrum of the synthetic utterance will be very similar to that of the original. But the method is essentially a time domain technique, hence its inclusion in this chapter.

5.2 Linear prediction

LPC of speech (Itakura & Saito, 1968, 1970; Atal, 1970; Atal & Hanauer, 1971; Makhoul, 1975*a*, *b*; Markel & Gray, 1976; Rabiner & Schafer, 1980), is a highly efficient technique whereby an utterance may be analysed to yield a set of parameters from which the original speech may be resynthesised. LPC has become a standard method for encoding speech signals in parametric form, for both transmission and storage, at low bit rates. The method is completely automatic and does not require formant tracking or spectral fitting; it is therefore ideal for vocoder applications –

though it does have a high processing overhead and is difficult to implement in real-time (Goldberg & Shaffer, 1975). The analysis and resynthesis procedures in LPC are intimately related to each other and are, in effect, mirror images.

In order to explain how LPC is accomplished it is first necessary to describe the idea of a linear predictor. Fig. 5.3 illustrates how a sampled data signal $S(n)$ may be predicted by forming a linear weighted sum of its past values. The predicted value $S_p(n)$ at the time increment n, is given by

$$S_p(n) = a_1 S(n-1) + a_2 S(n-2) + \cdots + a_i S(n-i) + \cdots + a_p S(n-p),$$

where $a_1, a_2, \ldots a_p$, are the predictor coefficients. The predicted signal, $S_p(n)$, may or may not be a good estimate of $S(n)$, the true value. However, it is necessary to assume initially that $S(n)$ can be predicted in this way with reasonable accuracy. It is then possible to examine the 'prediction error', defined by

$$e(n) = S(n) - S_p(n)$$

and to determine ways of making it a minimum. Fig. 5.3 illustrates the way in which $e(n)$ may be derived. Obviously, at any given time increment n, the value of $e(n)$ can be made equal to zero by choosing an appropriate set of predictor coefficients, a_i, $i = 1$ to p. But to minimise $e(n)$ on a sample-by-sample basis is not of any real advantage since the prediction coefficients will have to change at the same rate. The great efficiency of LPC is that it enables the prediction coefficients to be chosen so that $e(n)$ is optimally small over a large number of samples $(M+1)$, where $M \gg p$. This means that $(M+1)$ signal samples are effectively defined by a small set of numbers

Fig. 5.3. Linear prediction of speech. Schematic representation of prediction scheme, and generation of error singal $e(n)$.

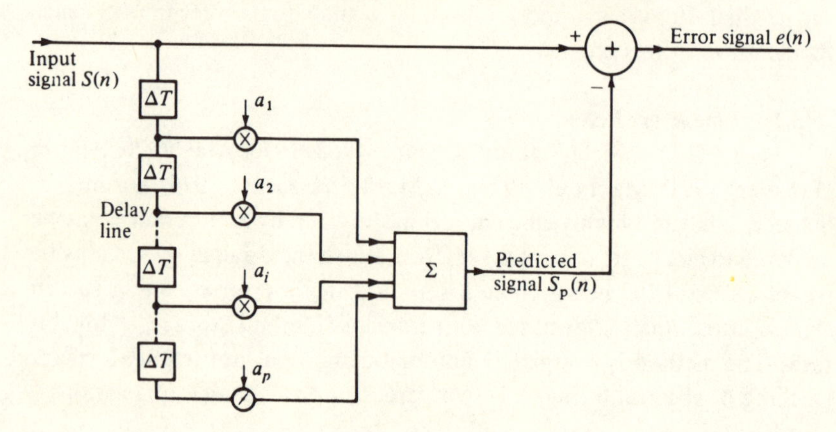

p (the predictor coefficients). The LPC algorithm minimises the energy in $e(n)$ over the interval $n = 0$ to $n = M$. That is, the prediction error squared and summed over $M + 1$ samples

$$E_n = \sum_{n=0}^{M} e^2(n)$$

is minimised by selection of suitable predictor coefficients a_i. This mean-squared-error minimisation will be discussed in some detail, but before doing so it is worthwhile pausing to consider the exact nature of the error signal itself (Strube, 1974; Rabiner, Atal & Sambur, 1977).

To say that the output signal of a given system is predictable implies that it has been produced by some systematic process. Furthermore, a procedure for predicting it would have to reflect in some way the structure of the originating system. It is fortunate that speech, most of the time (non-nasal voiced sounds), originates in a system which has an easily predicted output. This system has an all-pole transfer function which enables the predictive structure of Fig. 5.3 to predict its present output from past outputs. To see why this is so, it is convenient to consider the way in which the speech may be resynthesised from its LPC parameters. This procedure is shown diagrammatically in Fig. 5.4. Here the original speech signal is regenerated from the error signal $e(n)$ and the predictor coefficients a_i. That is,

$$S(n) = e(n) + \sum_{i=1}^{p} a_i S(n-i)$$

Fig. 5.4. Idealised LPC resynthesis. If $e(n)$ is used as the input signal, then $S(n)$ is regenerated exactly.

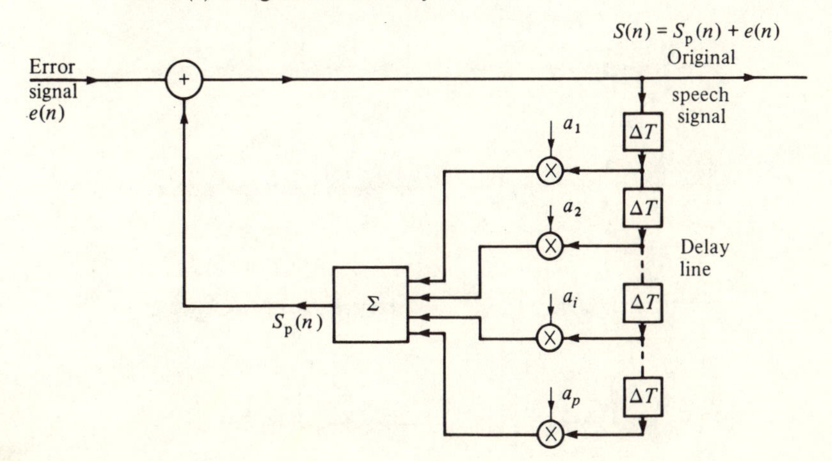

Now this expression may be written in z-transform notation,

$$SS(z) = E(z) + \sum_{i=1}^{p} a_i z^{-i} SS(z)$$

so that

$$SS(z) = \frac{E(z)}{1 - \sum_{i=1}^{p} a_i z^{-i}} = E(z)H(z)$$

where $SS(z)$ and $E(z)$ are the z-transforms of $S(n)$ and $e(n)$, respectively. That is, the output speech signal $S(n)$ is the result of an all-pole digital filter $H(z)$ being excited by an input function $e(n)$. Clearly, this structure may be viewed as a terminal-analogue synthesiser in which $H(z)$ represents the vocal tract transfer function and $e(n)$ is the excitation signal. That an all-pole filter is a good model for non-nasalised, voiced speech has already been shown in chapter 3. Thus, analysis using a linear predictor is exactly right for this kind of speech signal, and the predictor coefficients will be meaningful in that they will describe the system in which the speech is

Fig. 5.5. Differentiated sound pressure (left), and prediction error (right), for vowels /i, e, a, o, u, y/. Length of segment is 34 ms, sampling frequency is 8.825 kHz, prediction is twelfth order with a Hamming window. From Strube (1974).

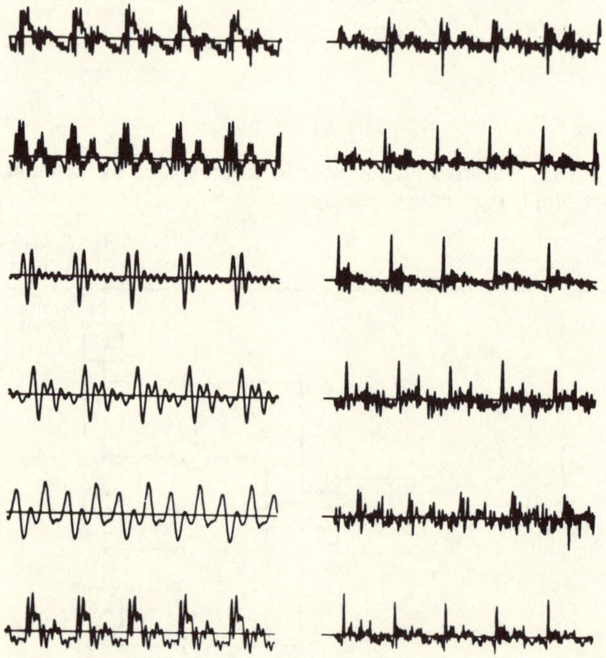

assumed to have originated, that is the vocal tract. Moreover, since vocal tract shape can only change relatively slowly, the predictor coefficients need only be updated at a moderate rate, typically every 10 mS (Chandra & Lin, 1974, 1977).

An examination of the exact waveform of the error signal $e(n)$ reveals the essential subtleties of LPC analysis. In the resynthesis structure of Fig. 5.4, $e(n)$ acts as an excitation signal for an all-pole terminal-analogue of the vocal tract. Thus for voiced speech $e(n)$ ought to be periodic at pitch frequency, and for unvoiced speech it ought to resemble random noise. Fig. 5.5 shows examples of $e(n)$ taken from Strube (1974), for various vowel sounds. It can be seen that $e(n)$ is quite small except at the moment of glottal excitation, when it jumps to a relatively high value. This is a consequence of the fact that the system response after excitation can be predicted quite well by an all-pole model, whereas the actual instant of glottal excitation is itself inherently unpredictable.

In the case of unvoiced speech minimising E_n has the effect of making the waveform of $e(n)$ look like random noise, that is, its spectrum is almost flat (Markel, 1972*a*). The 'predictability', in the form of spectral shaping, has been removed by the LPC algorithm, and is encoded in the predictor coefficients. For voiced speech (that is, periodic excitation), minimising E_n makes the spectrum of $e(n)$ only approximately flat. Because $e(n)$ is periodic, its spectrum must be harmonic, and it is the envelope of this harmonic structure that is flattened. This correlates with the fact that the waveform of $e(n)$ is approximately an impulse train at pitch frequency, as shown above. The actual shape of the glottal pulse is taken care of by the spectral flattening process of the LPC algorithm. Thus, if the glottal pulse shape is considered to be due to a hypothetical 'glottal filter', then this filter is included in the LPC system function $H(z)$. This may or may not be convenient depending on the application. In general it is not necessary to separate the glottal-shape from the vocal-tract component.

The great utility of LPC analysis and resynthesis is due to the form of the error function $e(n)$. In the case of voiced speech the waveform of $e(n)$ is approximately an impulse train at pitch frequency. For resynthesis $e(n)$ may be 'stylised' as a true unit impulse train, and only its frequency and amplitude need be encoded. For unvoiced speech, $e(n)$ is random noise and it is not necessary to reproduce its waveform exactly. For resynthesis it is permissible to stylise $e(n)$ as white noise and only its amplitude need be encoded (Markel, 1972*b*; Strube, 1974). Fig. 5.6 shows the arrangement of an LPC synthesiser using stylised excitation, as proposed by Atal & Hanauer (1971). The excitation control data consists of one bit for the voiced/unvoiced decision, and 6 bits for the periodic frequency. The amplitude of the speech output is controlled by the amplifier G, specified by 5 bits. The predictor coefficients require a total of 60 bits (12×5). Thus 72 bits per frame are required to characterise the input speech. This particular system is pitch synchronous, so that one frame of data is supplied per pitch period for voiced speech, and one frame per 10 ms for unvoiced speech. The average data rate for a male speaker is thus about 5000 b/s – a considerable saving over the usual 64 000 b/s required for Pulse Code Modulation (PCM).

One of the main factors affecting data rate of LPC parameters is p, the

Fig. 5.6. Practical LPC synthesiser. The excitation $e(n)$ is replaced by stylised sources; unit r.m.s. random noise, or a unit impulse train at pitch frequency. Amplitude is controlled by G, gain of amplifier.

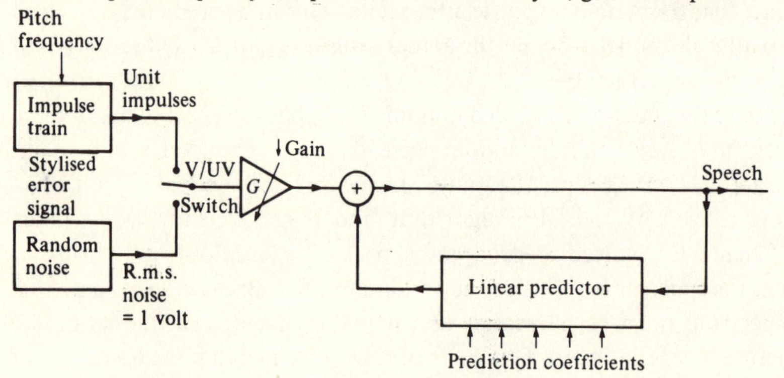

Fig. 5.7. Normalised r.m.s. prediction error, plotted against p, the prediction order, for voiced and unvoiced speech. From Atal & Hanauer (1971).

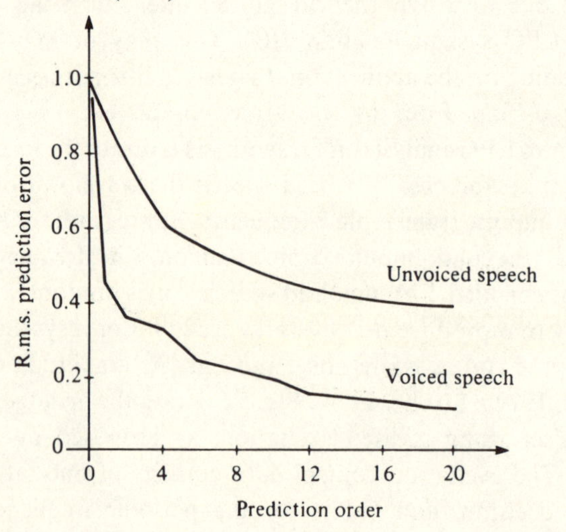

number of prediction coefficients. On theoretical grounds, to encode the effects of five formants would require five complex conjugate pole pairs, that is, ten prediction coefficients. Also, if the shape of the glottal pulse is to be accounted for by $H(z)$, then an extra two poles will be required to give a spectral fall-off proportional to frequency squared. This suggests that $p = 12$ is a reasonable estimate of the prediction order. Fig. 5.7 shows a graph of rms prediction error (Atal & Hanauer, 1971), plotted against p, the prediction order, for both voiced and unvoiced speech. Both graphs display a monotonic fall-off of r.m.s. error as p increases. But clearly, it is a case of diminishing returns. As expected, $p = 12$ is a reasonable estimate for voiced speech. For unvoiced speech the error is larger because of the greater randomness of the excitation. However, $p = 6$ seems to be a reasonable break point. This lower value of p for unvoiced speech is probably due to the fact that fricatives have only two or three formants.

5.3 Calculating predictor coefficients

It is now necessary to show exactly how an optimum set of predictor coefficients a_i, $i = 1, 2, \ldots, p$, may be calculated for a given segment of speech. The strategy to be used is that the average square error will be minimised over a time interval equivalent to $(M + 1)$ samples. That is, the function

$$E_n = \sum_{n=0}^{M} e^2(n)$$

is to be minimised by simultaneous variation of the predictor coefficients a_i. Since

$$e(n) = S(n) - S_p(n)$$

$$e(n) = S(n) - \sum_{i=1}^{p} a_i S(n-i)$$

the actual function to be minimised is

$$E_n = \sum_{n=0}^{M} \left[S(n) - \sum_{i=1}^{p} a_i S(n-i) \right]^2$$

This is done by the usual method of setting the partial derivatives of E_n with respect to the a_i simultaneously equal to zero. This must be done p times, that is once for each of the a_i, $i = 1, 2, \ldots, p$. To avoid confusion with the a_i in the summation, the stepped variable for the partial differentiation will be denoted by k. Thus,

$$\frac{\partial E_n}{\partial a_k} = 0, \quad k = 1, 2, \ldots p$$

This gives p simultaneous equations with p unknowns. Thus the set of equations is

$$\sum_{n=0}^{M} 2\left[S(n) - \sum_{i=1}^{p} a_i S(n-j) \right][-S(n-k)] = 0, \quad k = 1, 2, \ldots, p$$

By changing the order of summation each equation can be rewritten

$$\sum_{n=0}^{M} S(n-k)S(n) = \sum_{i=1}^{p} a_i \sum_{n=0}^{M} S(n-k)S(n-i), \quad k = 1, 2, \ldots, p$$

This system of equations is linear and therefore a solution is possible by matrix inversion. However, before attempting this it is worthwhile pausing for a moment to investigate the meaning of the summation terms,

$$\sum_{n=0}^{M} S(n-k)S(n) \quad \text{and} \quad \sum_{n=0}^{M} S(n-k)S(n-i)$$

These are similar in form to autocorrelation functions except that the limits of summation are not plus and minus infinity. They can, however, be viewed as short-time autocorrelation functions, if the speech segment is 'windowed' to make $S(n)$ equal to zero outside the interval $n = 0$ to M. An example of a windowing function is illustrated in Fig. 5.8. Since $S(n)$ is now zero outside the summation limits, the term

$$\sum_{n=0}^{M} S(n-k)S(n)$$

Fig. 5.8. Windowing of speech segment for auto-correlation method of calculating LPC coefficients.

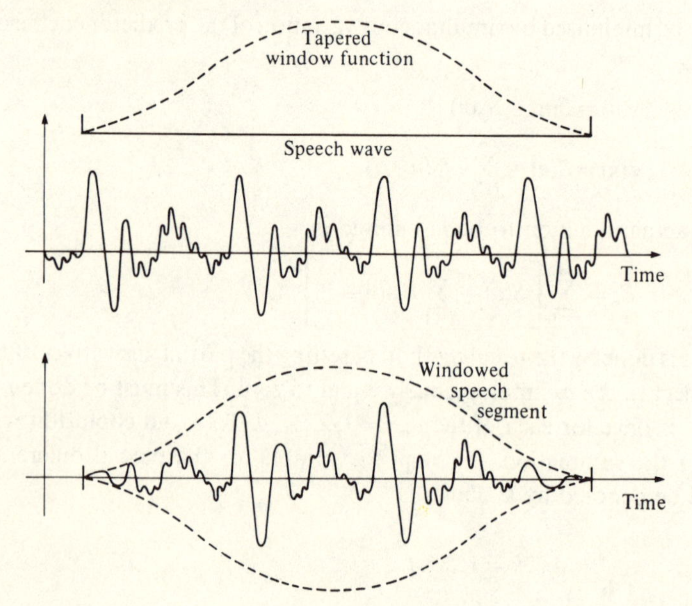

is the short-time autocorrelation function of the speech segment with a lag of k samples. That is,

$$\sum_{n=0}^{M} S(n-k)S(n) = R(k)$$

By similar argument and shifting of the limits, the term

$$\sum_{n=0}^{M} S(n-k)S(n-i) = \sum_{n=k}^{M+k} S(n)S(n+k-i) = R(i-k)$$

The system of equations can now be expressed more concisely as

$$R(k) = \sum_{i=1}^{p} a_i R(i-k), \quad k = 1, 2, \ldots, p$$

or in matrix form

$$[R(k)] = [R(i-k)][a_i]$$

Since $R(i-k) = R(k-i)$, the matrix $[R(i-k)]$ is symmetrical; also all elements on a given diagonal ($i-k = \text{constant}$) will be equal. This special form of matrix is known as 'Toeplitz', and the inversion of $[R(i-k)]$ is thereby simplified. This method of interpreting the summation limits of the equations is known as the autocorrelation method (Markel & Gray, 1973).

An alternative form of solution is to take the summation terms as they stand and evaluate them without windowing the speech segment. The terms are then no longer autocorrelation functions but they do have similar properties. Let the term

$$\sum_{n=0}^{M} S(n-i)S(n-k) = Q(k,i)$$

The term

$$\sum_{n=0}^{M} S(n-k)S(n)$$

is equivalent to $Q(k,i)$ with $i = 0$, that is $Q(k,0)$.

The system of equations may now be written in matrix form

$$[Q(k,0)] = [Q(k,i)][a_i], \quad k = 1, 2, \ldots, p$$

Again, $Q(k,i) = Q(i,k)$, so that the matrix is symmetrical, but elements on a given diagonal are not equal. A solution can be found by matrix inversion, but in this case the inversion process is not simplified by the Toeplitz condition. However a compensating factor is that there is now no need to multiply the speech segment by a window function. The matrix $[Q(k,i)]$ is known as the covariance matrix, and this method of finding the prediction coefficients is known as the 'covariance method' (Morf *et al.*, 1977).

5.4 Resynthesis problems

One problem encountered in resynthesising speech with the system shown in Fig. 5.6 is that of determining the gain parameter G used to restore the amplitude of the speech to its correct level. In the idealised structure of Fig. 5.4 this parameter was unnecessary, since the amplitude information was contained in the excitation (error) signal $e(n)$. In the practical system of Fig. 5.6 the excitation is stylised as either periodic impulses of unit amplitude, or random noise of unit r.m.s. value. An obvious way of evaluating the gain parameter G would be to calculate the r.m.s. value of $e(n)$ over the interval $n = 0$ to $n = M$. That is

$$G^2 = \sum_{n=0}^{M} e^2(n) = E_n$$

This is a simple enough relationship; the difficulty is that though E_n is minimised in the calculation of the a_i, in some methods of solution E_n is never actually calculated. A quicker way to estimate G has been suggested by Rabiner & Schafer (1980). It relies on the relationship between E_n, the prediction coefficients and the autocorrelation functions. It is

$$E_n = R(0) - \sum_{i=1}^{p} a_i R(i)$$

Thus G can be found from information already available in finding the coefficients.

Another problem, much more fundamental, lies in the form of the resynthesis function

$$H(z) = \frac{1}{1 - \sum_{i=1}^{p} a_i z^{-i}}$$

To resynthesise the speech using this equation as it stands requires great accuracy in specifying the values of a_i. This can readily be appreciated by noting that the resonances of $H(z)$ (its maximum values) are the result of the denominator's becoming small, and for this to occur the elements of the summation must cancel each other to a large extent. Thus small errors in

Fig. 5.9. Alternative method of realising prediction filter, as a cascade of second order functions.

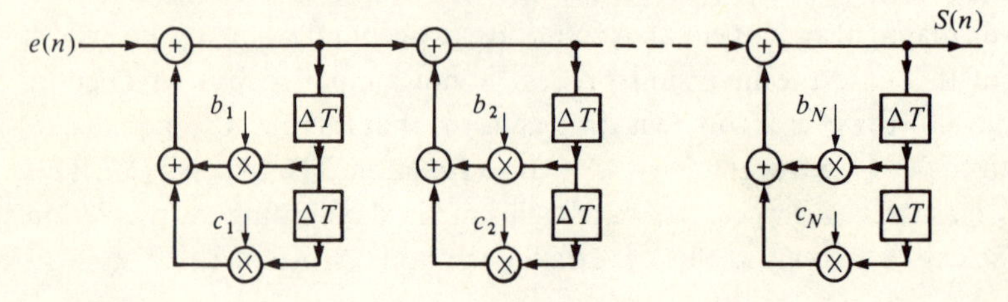

the a_i may lead to large errors in $H(z)$ – especially at the maxima. This is the well known sensitivity problem in the realisation of digital filters. It is usually solved (Atal & Hanauer, 1971) by factorising the denominator function to give a product of complex conjugate pole pairs, that is,

$$1 - \sum_{i=1}^{p} a_i z^{-i} = \prod_{k=1}^{p/2} (1 + b_k z^{-1} + c_k z^{-2})$$

$H(z)$ is then realised as a cascade of second order functions as shown in Fig. 5.9. The sensitivity problem can be reduced further if the second order coefficients are factorised into pole-frequency and bandwidth parameters. However, transforming the prediction coefficients into other, less sensitive parameters only alleviates the storage and transmission problem. The values of the coefficients must still be calculated to a high degree of accuracy before they can be transformed to a more convenient form. Moreover, finding the roots of the polynomial

$$1 - \sum_{i=1}^{p} a_i z^{-i}$$

is not a trivial problem.

A much more efficient, and altogether more satisfactory way of finding insensitive LPC parameters has been proposed by Itakura & Saito (1970, 1972), which leads to a completely different synthesis structure – known as a lattice network. Alternative formulations of this same basic technique are given by Morgan (1975) and Burg (1975).

5.5 Lattice LPC – PARCOR parameters

The linear prediction scheme of Itakura & Saiko (1970, 1972) is illustrated in Fig. 5.10. In this analysis structure there are two error signals which progress through the network and are optimised at each stage to give the k-parameter of that particular section. The two error signals at the ith stage and for the nth sample are the forward error $e_i^+(n)$ and the backward

Fig. 5.10. Schematic arrangement of lattice LPC. $k_1 \ldots k_i \ldots k_P$ are partial correlation coefficients. After Itakura & Saito (1970).

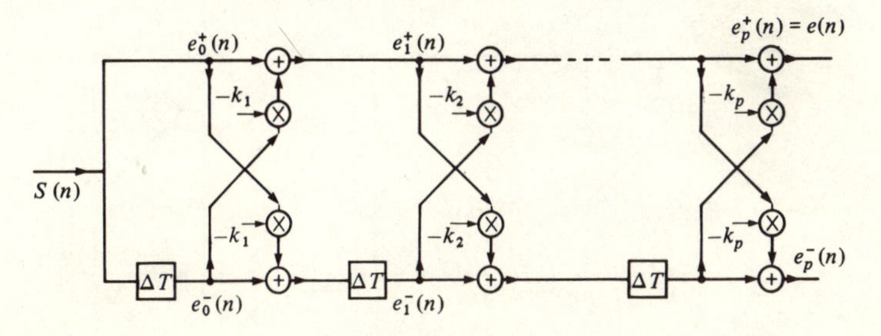

error, $e_i^-(n)$. The coefficient k is calculated from these two error signals over a suitable interval, $(M+1)$ samples, by the formula

$$k_{i+1} = \frac{\sum_{n=0}^{M} e_i^+(n) e_i^-(n)}{[\sum_{n=0}^{M} e_i^+(n)^2 \cdot \sum_{n=0}^{M} e_i^-(n)^2]^{1/2}}$$

The coefficient k_1 is calculated from $e_0^+(n)$, which is the input signal $S(n)$, and $e_0^-(n)$, which is $S(n-1)$. The k_1 coefficient is thus, in effect, a linear prediction of order one. The cross-coupling of the lattice then uses k_1 to remove this first order predictability from the signal, to give the error signals $e_1^+(n)$, and $e_1^-(n)$. The relationship is

$$e_1^+(n) = e_0^+(n) - k_1 e_0^-(n-1)$$

and

$$e_1^-(n) = e_0^-(n-1) - k_1 e_0^+(n)$$

This procedure is repeated at each stage and the set of k_i coefficients is extracted recursively. Since these coefficients are calculated from a partial correlation of forward and backward errors, these parameters are known as PARCOR coefficients. The partial correlation is actually normalised so that the k-parameters are within the range -1 to $+1$. In the resynthesis structure, shown in Fig. 5.11, the transfer function is relatively insensitive to inaccuracies in the k-parameters, and PARCOR coefficients can be used in a practical synthesiser without modification. The synthesis lattice of Fig. 5.11 is the complement of the analysis structure of Fig. 5.10. As before, the forward and backward error signals propagate through the network. Ideally, the input to the synthesis lattice is the final error signal from the analysis lattice $e(n)$ (sometimes known as the residual). In practice, $e_p^+(n)$, as before, is stylised to be either random noise or an impulse train at pitch frequency.

In the synthesis structure of Fig. 5.11, the amplifier gain parameter G is introduced to restore the signal amplitude to its correct level. In the Itakura

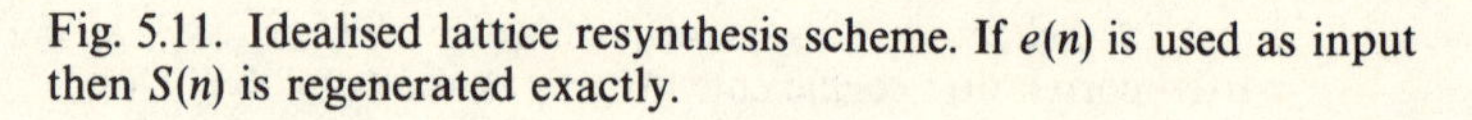

Fig. 5.11. Idealised lattice resynthesis scheme. If $e(n)$ is used as input then $S(n)$ is regenerated exactly.

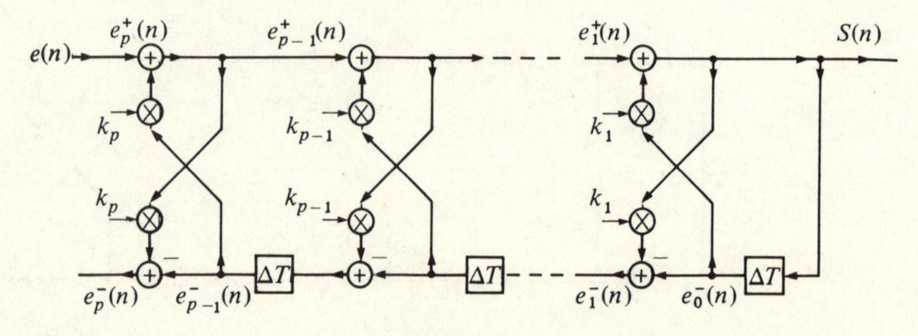

scheme (Itakura & Saito, 1970), G is calculated from the final error signal (residual), by the relation

$$G^2 = \sum_{n=0}^{M} e_p{}^+(n)^2$$

Autocorrelation of the residual is used to give a voiced/unvoiced decision and to estimate the pitch period.

The elegance and simplicity of this LPC method makes it very attractive and convenient to implement (Samei, Asada & Nakata, 1980). In practice, alternative synthesis structures are used to minimise the number of multiplications required. Fig. 5.12 shows a lattice section which requires only one multiplication.

The synthesis structures of Fig. 5.11 and Fig. 5.12 are somewhat reminiscent of the digital delay line model of the vocal tract discussed in chapter 3. Also, the idea of forward and backward error signals is very similar to the concept of incident and reflected waves in the delay line. It can be shown that there is, in fact, a direct relationship between the two formulations. The input–output relationship for the ith section of the lattice of Fig. 5.11 can be shown to be

$$\begin{bmatrix} e_i{}^+ \\ e_i{}^- \end{bmatrix} = \begin{bmatrix} 1 & -k_i z^{-1} \\ -k_i & z^{-1} \end{bmatrix} \begin{bmatrix} e_{i-1}{}^+ \\ e_{i-1}{}^- \end{bmatrix}$$

Similarly, for the ith section of a digital delay line, the incident and reflected waves into and out of the section are given by

$$\begin{bmatrix} P_i{}^+ \\ P_i{}^- \end{bmatrix} = \frac{1}{1+r_i} \begin{bmatrix} 1 & -r_i z^{-1} \\ -r_i & z^{-} \end{bmatrix} \begin{bmatrix} P_{i+1}{}^+ \\ P_{i+1}{}^- \end{bmatrix}$$

where r_i is the reflection coefficient of the section. The form of these two

Fig. 5.12. Alternative lattice section, using only one multiplication.

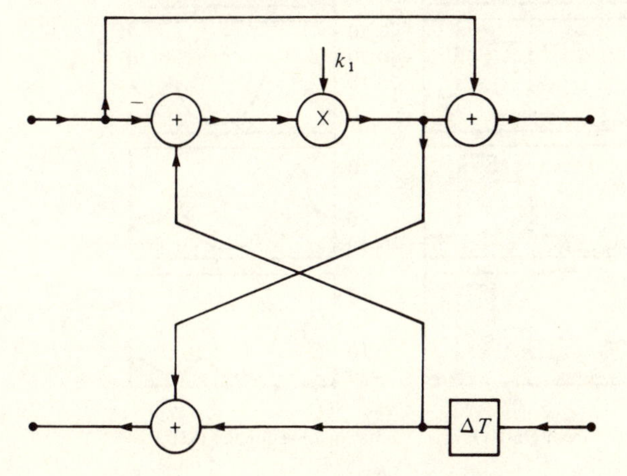

expressions is the same except for the factor $1/(1+r_i)$.

The transfer function of the whole digital line, or lattice filter, may be obtained by multiplication of the matrices representing the cascade of sections. Since the form of the matrices is the same in each case, the resulting transfer functions will be the same, except for the factor

$$\prod_{i=1}^{p} \frac{1}{1+r_i}$$

Since this factor does not contain the delay variable z^{-1}, the transfer function polynomials in z will be the same for each structure, if the k_i are made equal to the r_i.

This interesting relationship can be used to obtain vocal tract area functions directly from the speech signal. Wakita (1973) has shown that if the speech signal is properly pre-emphasised, and the glottis and lips terminated correctly, then quite reasonable vocal tract shapes can be obtained. Since the reflection coefficients give only ratios of areas of adjacent sections, and

$$r_i = \frac{A(i+1) - A(i)}{A(i+1) + A(i)}$$

Fig. 5.13. Vocal tract area functions and transfer functions obtained for five American vowels. From Wakita (1973).

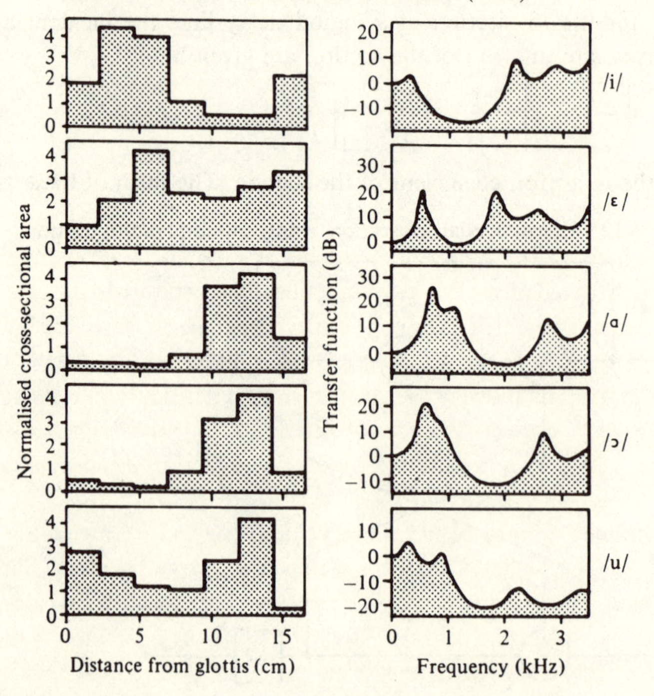

it is necessary to know one area absolutely before the other areas can be found. Wakita takes the glottal area to be fixed and from this calculates the other areas recursively, using the relation

$$A(i) = \frac{1 - r_i}{1 + r_i} A(i+1)$$

An example is given in Fig. 5.13.

One of the main motives for using PARCOR coefficients is that they are less sensitive to errors and thereby may be quantised to fewer bits than conventional predictor coefficients (Makhoul, 1977). Viswanathan & Makhoul (1975) have examined the sensitivity of LPC parameters derived by various methods. They have shown that the parameters which have optimum sensitivity are related to PARCOR coefficients by the formula

$$p(i) = \ln\left(\frac{1 + k_i}{1 - k_i}\right) = \ln\left(\frac{A(i+1)}{A(i)}\right)$$

Since this ratio of the k_i parameters is, in effect, a ratio of vocal tract areas, these optimum parameters are known as 'log area ratios' (Strube, 1977).

5.6 Inverse filtering

Linear prediction of speech signals can also be thought of as a process of 'inverse' filtering. The arrangement of Fig. 5.3, in which the predicted signal is subtracted from the original, is, in effect, an inverse filter. The digital filtering function is simply

$$1 - \sum_{i=1}^{p} a_i z^{-i}$$

that is, an all-zero filter.

Since the speech signal is assumed to have originated in an all-pole system, this all-zero filter (defined by the linear prediction) is simply the inverse of the assumed all-pole system function. Thus, the error signal $e(n)$, which is the result of inverse filtering the speech signal, is clearly the input signal to the assumed system. Again, it must be noted that the effect of the glottal pulse shape is included in the vocal transfer function, $H(z)$. Hence the excitation function derived by inverse filtering the speech signal is an impulse train, rather than an approximation to real glottal pulses.

The inverse filtering approach has been used by Markel (1973) to investigate the behaviour of formants in speech. Using an LPC algorithm to derive a suitable inverse filter, Markel obtained the error (excitation) signal from a segment of voiced speech. The spectrum of the original speech segment and the spectrum of the LPC error are shown in Figs. 5.14(*a*) and (*b*) respectively. Both spectra are harmonic at pitch frequency; it is however, that the error spectrum is a flattened version of the original speech spectrum. Fig. 5.14(c) shows the frequency response of the 'inverse' filter which has performed the flattening. The reciprocal of the all-zero, inverse filter, is an all-pole filter $H(z)$, which is the system function of the 'assumed system'. The frequency response of $H(z)$ is shown in Fig. 5.14(d).

Markel (1973) has shown that a simple peak-picking routine operating

Fig. 5.14. (a) FFT (Fast Fourier Transform) spectrum of voiced speech segment. (b) FFT spectrum of LPC error. (c) Frequency response of all-zero 'inverse' filter (d) Frequency response of all-pole filter $H(z)$. From Markel (1972a).

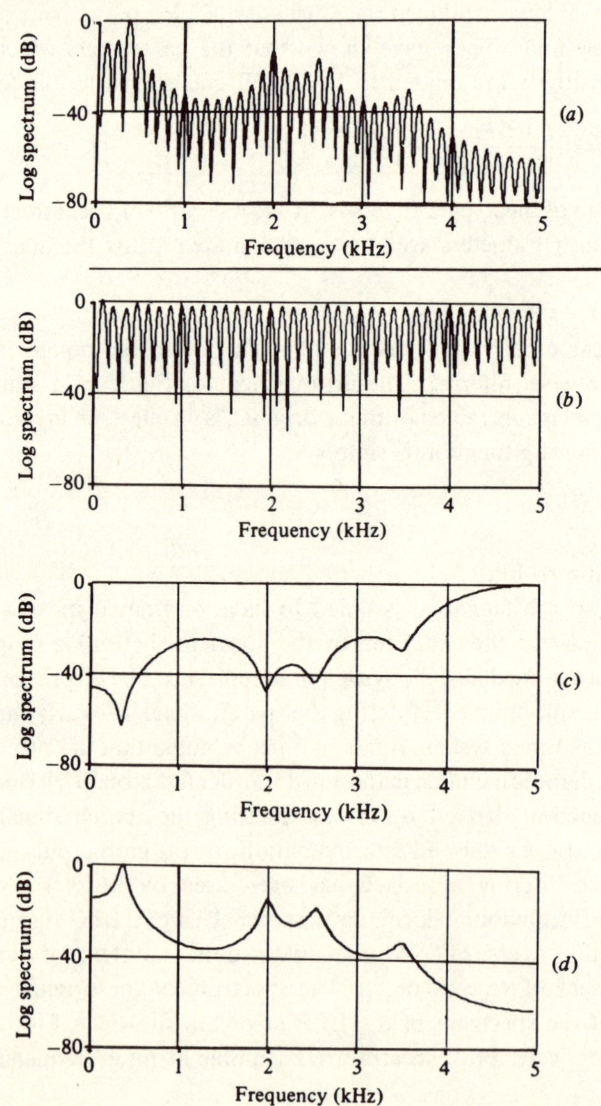

on the frequency response of $H(z)$ is sufficient to extract formants for most examples of voiced speech. Optimum conditions for speech sampled at 10kHz are $p = 14$, using a Hamming window on a time aperture of 20–30 ms. McCandless (1974) has developed a simple continuity algorithm to track formants in regions of confusion.

5.7 Conclusions

For non-nasal, voiced sounds an all-pole model of speech production is very satisfactory, as has been shown in chapters 3 and 4. For LPC derived parameters, both pole frequencies and bandwidths are extracted to give very good approximations to the vocal tract glottal filter functions. These parameters can even be used to compute vocal tract area functions which agree reasonably well with X-ray data. However, for nasals, stops and fricatives, it would be advantageous to be able to include spectral zeros in the vocal tract transfer function.

It is possible to approximate a zero term $(1 - az^{-1})$ by the expansion

$$\frac{1}{1 + az^{-1} + a^2z^{-2} + a^3z^{-3} + \cdots + a^Nz^{-N}}$$

This is an all-pole function, which becomes more accurate as N tends to infinity. In fact this is how the LPC algorithm manages to fit signals which originate in systems which have zeros as well as poles. Thus, LPC always gives the best possible all-pole fit even when an all-pole model is not appropriate. This can be viewed as an advantage if all that is required is straightforward parameterisation of the speech signal. It is, however, a disadvantage if it is hoped to use the LPC model to derive vocal tract area functions for other than non-nasalised voiced speech. Because in this case the vocal tract system will contain zeros, and the LPC algorithm will approximate each zero by a set of poles, which will obscure the values of the true poles of the system.

Attempts to include spectral zeros in the LPC model (Longuet-Higgins, 1978) have not been entirely successful. The reasons for this are not quite clear, but it seems as if the zeros 'compete' with the poles to represent the signal, and the exact division of the task varies from frame to frame in an unstable manner.

To summarise, LPC is an extemely efficient way of parameterising speech signals for low bit rate transmission or storage. While some formant synthesisers can produce synthetic speech with better quality, they generally require some hand crafting of the control signals to do so, whereas, LPC parameters are derived completely automatically, quickly, and with the guarantee that they are the best possible all-pole fit.

6 *Synthesising control parameters*

The art of producing intelligible, natural-sounding synthetic speech

'Soon I deduced... that the sounds of speech are only truly distinguished by virtue of the proportion that exists between them, and that they obtain perfect clarity only through the relationship established between whole words and phrases.'
'What seems certain is that one could not operate such an instrument [talking machine] in accordance with theoretical calculations: one would always have to work out the statistics by practical research...'
Von Kempelen, 1791

6.1 Introduction

In order to synthesise speech-like sounds, it is necessary to supply a synthesiser with a sequence of control signals which are appropriate to both the details of the required utterance and the design of the synthesiser. The intelligibility and naturalness of the resultant speech will depend, to a large extent, on the quality of the control parameters, and only marginally on the particular type of the synthesiser being used. The problem of designing adequate synthesisers may well be said to have been solved; the more difficult and interesting challenge is to devise ways of specifying synthesiser control signals, to produce intelligible and natural-sounding speech.

One obvious way of obtaining control parameters for a synthesiser is to derive them from analysis of real speech. Thus, it is possible to trace out formant trajectories from spectrograms, and use them as control signals for a formant synthesiser. Additional information concerning voicing, frication and amplitude are easily obtained, and included in the control data. The result is (usually) an intelligible, though machine-like, replication of the original utterance. Using a more scientific approach, speech can be subjected to LPC analysis, and the resultant parameters used to control a matched LPC synthesiser, as described in chapter 5. Here the resynthesised utterance may be almost indistinguishable from the original, depending on the order of the prediction, and the level of parameter quantisation.

However, whatever method is used to extract parameters from real speech, the only advantage of this kind of synthesis over direct recording of the original speech wave is that of information economy. It is, therefore, of great utility in the encoding of speech, for transmission at low bit rates (vocoders), and in applications where a vocabulary of standard words and fixed phrases are required to be stored (for example, in telephone information services). But for speech output from computers or other automatic systems, where a large or unlimited vocabulary is required, parameterised real speech is not really flexible enough, especially if the speech is to be connected and continuous.

The alternative method of generating control signals for a synthesiser is to create them from a written specification of the required utterance. At the simplest level, such a specification might be prepared by an expert phonetician, using conventional IPA symbols plus some stress information. The synthesis problem is then to convert this symbol string into a suitable sequence of control signals. The solution is particularly straightforward if the synthesiser is an articulatory system model; in this case it is only necessary to define a set of articulatory control sequences, each corresponding to a particular phonetic symbol. Synthesis is then simply a concatenation of the control signals equivalent to the input phoneme string. If the dynamics of the articulatory model are correct, then co-articulation will take place automatically and realistically. On the other hand, if a signal model synthesiser is used, then co-articulation effects must be somehow incorporated into the control parameters themselves. This can be done, most conveniently, by using stored data concerning phoneme targets, and a set of rules which model co-articulatory behaviour.

One way of avoiding some of the problems of co-articulation is to use larger, multi-phonemic units as the 'building blocks' for constructing sentence length utterances. Both words and syllable-type units have been proposed for this purpose. The technique is to use a vocabulary of words or syllables, from which phrases can be built up by simple concatenation of selected units. The vocabulary is constructed by parametric storage of words or syllables taken from real utterances. This method combines the naturalness of parametric speech with the flexibility of phonemic synthesis, to provide output in real-time. Consequently, it has been used with success in several practical applications (Flanagan, Rabiner, Schafer & Denman, 1977; Fallside & Young, 1978).

At a more sophisticated level, it is possible to synthesise speech from orthographic text, that is, ordinary, written English. In this case it is necessary to use rules of a more sophisticated character, which are able to include the effects of prosodic stress, as well as word stress and the complexity of English spelling. Such 'supra-segmental' rules will be necessary for both the signal model, and the system model, types of synthesisers. These techniques not only require a large dictionary or lexicon, but need to have built-in knowledge of grammar and semantics.

Many different techniques have been proposed for the synthesis of speech and considerable research is still being applied to the problem (Flanagan, Coker, Rabiner, Schafer & Umeda, 1970). At present, speech synthesis from parametric storage (both formants and LPC) can give output indistinguishable from the natural utterances from which it was derived. However, speech synthesised from text (phonemic or orthographic), though very intelligible, is still machine-like in character. This reinforces the view, that the speech synthesisers themselves are perfectly adequate, and that refinements in the quality of synthetic speech must be sought through improvements in the synthesis of the control parameters. The uncertainty as to how this can be accomplished is reflected in the wide variety of techniques which are still under active research. The following descriptions and discussion, therefore, make no attempt to judge the issue, and are confined to as comprehensive a survey as is possible in the space available.

6.2 Vocoders

If a speech signal is sampled at an appropriate rate (8 kHz for 4 kHz bandwidth speech), and if each sample is quantised to an adequate number of levels (256 encodes to 8 bits), then the resulting sampled data signal may be stored or transmitted, digitally, at a data rate of 64 000 b/s. The vocoder (= voice coder) was specifically invented to take advantage of the redundancy and predictability of the speech waveform to encode speech at a data rate somewhat lower than this nominal 64 000 b/s.

The basic concept of the vocoder is based on the proposition formulated by H. Dudley (1940, 1955), that the speech signal is compounded of two parts, excitation and spectral envelope, and that it is possible to separate and recombine these parts, without loss of information. Moreover, since the variables associated with these parts change relatively slowly with time, they can be encoded at a much lower data rate than the original speech waveform. This technique, illustrated in Fig. 6.1, is effective for both transmission and storage of speech at low bit rates. The theoretical basis and validity of this implied source/system separation, has been discussed in chapter 3. In practice, the idea has been applied to a whole range of vocoding devices, which, over the years, have given more and more spectacular performance in the compression of speech data (Schroeder & David, 1960; Schroeder, 1966; Kelly, 1970).

The original machine invented by Dudley (1936, 1939*a*, *b*) is the channel vocoder, the synthesis part of which has been described in chapter 4. The complementary analyser uses a ten-channel filter bank to extract spectral amplitudes, and a special circuit to detect voicing information. Each of the eleven output parameters is then filtered to 25 Hz. So that, sampling each parameter 50 times per second, and encoding at 8 bits per sample, the overall data rate for the Dudley vocoder would have been about 4400 b/s, though it was never used for digital transmission. Later versions, optimised and digitally encoded, proved capable of data rates down to 2.4 kb/s (Gold & Rader, 1967).

An obvious development of the channel vocoder was to encode the spectral information as formant frequencies, leading to the formant vocoder (Coker, 1963, 1965). A major problem, in this machine, is the extraction of formants (Schroeder, 1956), which in unvoiced speech are not defined, and even in voiced speech are not always obvious. An example of a 'state-of-the-art' formant vocoder is the machine described by Chong (1978), which has a data rate of 1200 b/s. A difficulty which besets both channel and formant vocoders is the detection of voicing information. A solution to this problem has been to transmit a low pass filtered version of the speech signal (base-band), and use this to generate an excitation waveform. At the synthesiser, this baseband signal is used directly as the first formant output, and indirectly after processing, to excite the higher formants. Thus any ambiguity in the voicing is transmitted rather than resolved. Vocoders using this technique are known as voice-excited (Flanagan, 1960; Golden, 1963).

The problems of pitch detection and formant tracking have never been completely solved, and have been somewhat overtaken by the advent of LPC of speech signals (Itakura & Saito, 1970; Atal & Hanauer, 1971). In the LPC vocoder the formant information is contained in the LPC parameters, and though the resonances of the all-pole synthesis function may represent formants, it is not necessary to label them as such. The speech excitation information is contained in the LPC error signal (residual), which is 'stylised' before encoding. Thus, though a voiced/unvoiced decision has to

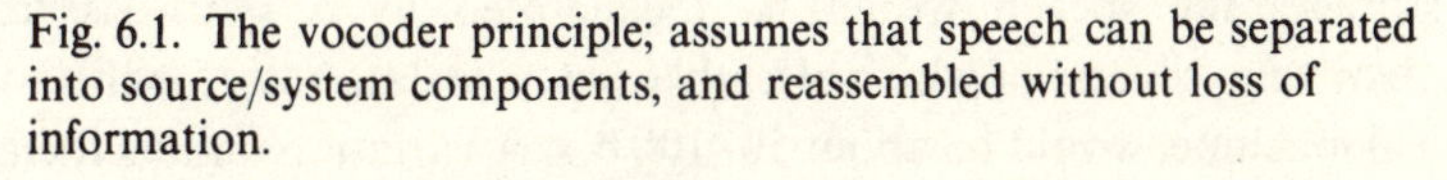

Fig. 6.1. The vocoder principle; assumes that speech can be separated into source/system components, and reassembled without loss of information.

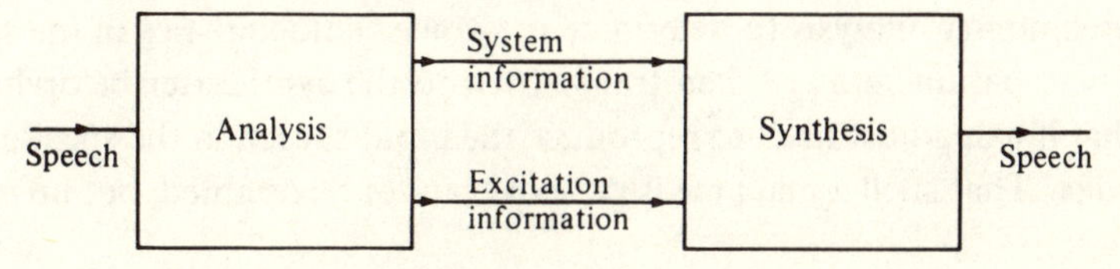

be made, it can change from frame to frame, and mixed or ambiguous excitation is reasonably well replicated. An interesting alternative is the 'residual-excited' vocoder (Chong & Magill, 1975), which uses the LPC error signal directly, to excite the LPC synthesiser, and gives excellent speech quality at 9·6 kb/s.

The LPC vocoder has many variants, particularly in the way in which voicing information is extracted. A wide range of data-compression/speech-quality trade-off are possible, from 4.8 kb/s (Goldberg and Ross, 1973) through 3.3 kb/s (Markel & Gray, 1974) to 2.4 kb/s (Hofstetter, Tierney & Wheeler, 1977). By encoding the LPC parameters themselves, using adaptive delta modulation, the data rate can be reduced as low as 1000 b/s (Sambur, 1975).

For applications where the time waveform of the speech signal is important, the LPC vocoder appears to be an optimum solution. However, if reproduction of speaker identity is not required, then data rate can be reduced even further, by use of the 'pattern matching' vocoder (Smith, 1963). In this machine the spectral information is also 'stylised', so that only a limited number of spectral shapes is transmitted. The input speech is analysed frame by frame, in the usual way, then each spectral frame is matched to the nearest standard spectral shape, by use of a pattern matching algorithm. The standard patterns are chosen to be typical of speech spectra, and the number of such patterns (typically 1000) is considerably less than the total number of shapes possible, from the spectral analysis. The resultant resynthesised speech is not of high quality; however it can be quite intelligible, and at a data rate of 500 b/s it represents a remarkable achievement in speech compression.

The ultimate vocoder would be one that used phonemic parameters. Though the analysis part of such a machine has not yet been invented, conceptually, its principle is quite valid, and can be simulated in experiments by a skilled phonetician. The essential idea is that the input speech is analysed to give a sequence of phonemic symbols plus stress markers. This information may then be further encoded as orthographic text, with conventional punctuation to imply the stress pattern. After transmission (or storage) the speech would be regenerated by a synthesis from text algorithm, several of which already exist. The data rate, for this hypothetical machine, would be about 50–100 b/s. A variation on this theme uses a preliminary analysis to determine the voice characteristics of the speaker. These parameters are then transmitted to the synthesiser beforehand, so that it can adjust itself to reproduce the input speech in the speaker's own voice. That such a machine is possible cannot be doubted, but how long it will be before it can be implemented, and whether or not it will be economic, remain to be seen.

6.3 Parametric storage of speech

The development of the vocoder was primarily intended to fulfil the need for bandwidth reduction in telephone systems. However, the same process which reduces the data rate for speech transmission can also be used for economical speech storage. With the spread of digital computers, this application of parametric speech has grown rapidly, permitting simple computer voice-response systems to be implemented (Rabiner & Schafer 1976; Itakura, Saito, Koike, Sawabeh & Nishikawa, 1972; Wiggins, 1978).

In the analysis half of a vocoder, the signal processing must be relatively straightforward to enable the parameters to be calculated in real-time. In the preparation of parameters for storage no such constraints apply, and since the computations may be carried out off-line, quite elaborate processing techniques may be employed. In fact, it is the advent of digital computers that has permitted the present high level of complexity in speech processing procedures. The result is that speech may be parameterised to very low data rates, whilst still retaining high intelligibility (Kun-Shan Lin & Ying L. Tsui, 1981).

The first such processing procedures were formant extraction algorithms, the outputs of which were designed to drive formant synthesisers. Original work in this field used formant trajectories copied from spectrograms of real speech, but this was a tedious, as well as inaccurate, method. Consequently, many automatic formant tracking algorithms have been proposed: Flanagan, 1956; Pinson, 1963; Suzuki, Kadokawa & Nakata, 1963; Schafer & Rabiner, 1970). None of these methods is perfect, though some are capable of producing parameters, from which speech may be resynthesised with good fidelity.

The problem with formants is that there may be too few or too many, or they may not exist at all. One way out of this dilemma is to copy the whole spectrum by a 'curve fitting' technique, whereby parameters are extracted to enable a given synthesiser to produce the best fit to the speech spectra. Spectrum fitting algorithms of this kind are not analytic and often require several time-consuming iterations (Bell, Fujisaki, Heinz, Stevens & House, 1961; Paul, House & Stevens, 1964; Seeviour, Holmes & Judd, 1976; Green, 1976).

The development of spectrum fitting lost impetus after the advent of LPC processing. Although LPC is intended to model the time waveform, it also gives a good fit in the spectral sense, which if not optimum is at least analytic. These methods have already been described in chapter 5, and in the above section on vocoders. In the preparation of LPC parameters for storage, however, two extra degrees of freedom are available. The first is in the choice of speaker – it has been shown that this can influence the acceptability of the resynthesised speech (Frantz & Kun-Shan Lin, 1982). The second is the amount of time available for processing: this is usually utilised to encode the LPC error signal more carefully (Wong & Markel, 1978; Sambur, Rosenberg, Rabiner & McGonegal, 1978; Atal & David, 1979). This often involves an iterative process, sometimes with human intervention.

The success of LPC parameters for speech storage and resynthesis is so great that almost all the commercially available parametric synthesisers use this technique, with manufacturers providing a custom parameter preparation service.

6.4 Synthesis from phonemic specification

The basic research on the correlation between perceived phonemic units and their acoustic patterns, as manifest in spectrograms, was carried out in the 1950s at the Haskins Laboratories (Cooper, Liberman & Borst, 1951; Cooper, Delattre, Liberman, Borst & Gerstman, 1959). This research showed that even though we perceive speech as a concatenation of discrete phonemes, at the acoustic level no such segmentation exists. Rather, individual phonemes appear to be very much 'overlapped' in their effect on the acoustic pattern. It would appear that continuously changing formant trajectories are best characterised as sequences of transitions between successive sets of 'target' frequencies, each set of target formants being the physical manifestation of a particular phoneme. The situation is complicated by the fact that, sometimes, the target formant frequencies are never reached. The culmination of this work was the proposal that speech should be synthesised by using rules to generate formant transitions between phoneme targets, abstracted from real speech and stored in look-up tables. These 'Minimum Rules for Speech Synthesis' (Liberman, Ingermann, Lisker, Delattre & Cooper, 1959) heralded the beginning of truly flexible synthetic speech from phonemic specification.

Since that time, several rule-synthesis algorithms, operating at the segmental level, have been proposed (Kelly & Gerstman, 1961; Estes, Kerby, Maxey & Walker, 1964; Holmes, Mattingly & Shearme, 1964; Rabiner, 1968*b*, 1969; Klatt, 1976*a*; Gagnon, 1978).

These synthesis methods vary in both the complexity of the synthesis rules, and the size of the phoneme look-up tables. At one extreme, the method of Estes *et al.* (1964) is to store actual transitions between phonemic

Fig. 6.2. Phoneme parameter tables for segmental synthesis from formant data. F_1, F_2 and F_3 are target formant frequencies, and R_1, R_2 and R_3 are their corresponding target regions. From Rabiner, 1968*b*.

Phoneme	F_1	R_1	F_2	R_2	F_3	R_3	Nasal	Fric	Voiced
Vowels									
i	270	75	2290	75	3010	150	–	–	+
ɪ	390	75	1990	75	2550	110	–	–	+
e	530	75	1840	80	2480	110	–	–	+
æ	660	75	1720	75	2410	110	–	–	+
ʌ	520	75	1190	75	2390	75	–	–	+
ɑ	730	37	1090	75	2440	115	–	–	+
ɔ	570	75	840	75	2410	115	–	–	+
ʊ	440	75	1020	75	2240	90	–	–	+
u	300	75	870	80	2240	90	–	–	+
ə	490	75	1350	80	1690	100	–	–	+
Semi-vowels									
w	300	25	610	40	2200	150	–	–	+
j	300	25	2200	110	3065	200	–	–	+
Liquids									
l	380	25	880	80	2575	150	–	–	+
r	420	30	1300	80	1600	100	–	–	+
Stops									
b	0	50	800	75	1750	120	–	–	+
d	0	30	1700	50	2600	160	–	–	+
g	0	15	2350	50	2000	100	–	–	+
p	0	50	800	40	1750	80	–	–	–
t	0	30	1700	30	2600	100	–	–	–
k	0	10	2350	30	2000	–	–	–	–
Nasals									
m	280	17	900	17	2200	40	+	–	+
n	280	17	1700	17	2600	100	+	–	+
ŋ	280	17	2300	17	2750	100	+	–	+
Fricatives									
f	175	30	900	50	2400	120	–	+	–
θ	200	30	1400	40	2200	100	–	+	–
s	200	30	1300	40	2500	70	–	+	–
ʃ	175	30	1800	100	2000	150	–	+	–
v	175	30	1100	50	2400	120	–	+	+
ð	200	30	1600	40	2200	100	–	+	+
z	200	30	1300	40	2500	70	–	+	+
ʒ	175	30	1800	100	2000	150	–	+	+

units, so that the synthesis rules simply select the appropriate 'diphones' and concatenate them to produce synthetic speech. The minimum number of such diphones required to synthesise the set of English phonemes is about 1000, and somewhat more for high quality speech (Dixon & Maxey, 1968).

On the other hand, the method of Rabiner (1968*b*, 1969) uses a very simple stored specification of each phoneme. His look-up tables, shown in Fig. 6.2, consist of three formant frequencies and their corresponding frequency regions, and excitation data. The rules, however, are somewhat complex. Transitions between formant targets are modelled by second order equations of motion, with critical damping. The time constants of these equations are different for each formant, and also depend on the phonemes involved in the transition. Starting from one set of target formant frequencies, the formant trajectories are calculated frame by frame, as they progress toward their new targets. As soon as all the formants are within the defined 'region' of their target values, new targets and their appropriate time constants are fed into the equations, and the calculations are continued. This is known as the 'base touching' technique. An exception to this procedure, is when the targets are those of a stressed vowel; then the targets are held for a specified duration before the next targets are fed in. Care is taken to ensure that the slopes of the formant trajectories are continuous, so that the method generates smooth transitions.

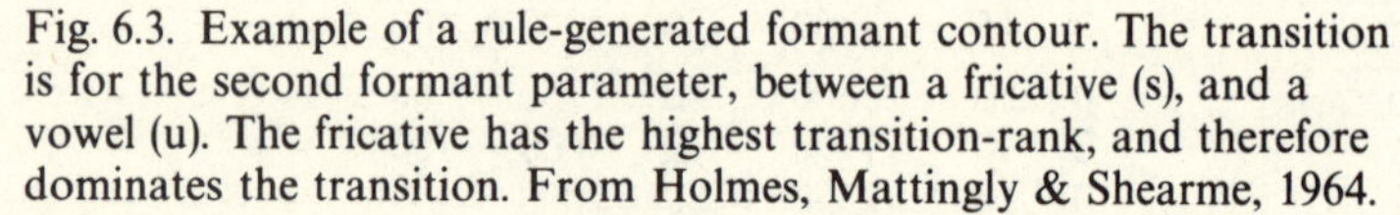

Fig. 6.3. Example of a rule-generated formant contour. The transition is for the second formant parameter, between a fricative (s), and a vowel (u). The fricative has the highest transition-rank, and therefore dominates the transition. From Holmes, Mattingly & Shearme, 1964.

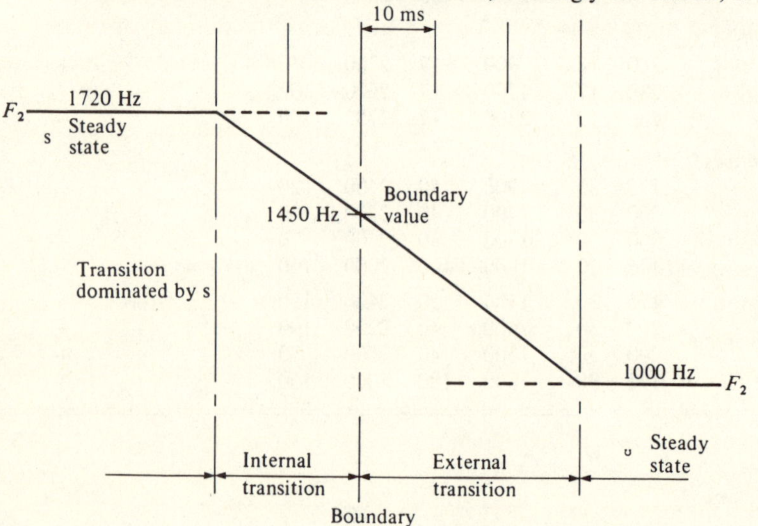

In the method due to Holmes *et al.* (1964), the phoneme look-up tables store three formant frequencies, four formant amplitudes, fixed and transitional duration data, excitation type and a parameter called the transition-rank. This factor determines which of two adjacent phonemes will dominate their mutual transition, and to what extent they will overlap. Stops have the highest rank, vowels the lowest. At the boundary between two phonemes, a 'boundary value' is calculated for each control parameter, using specified fractions of the parameter values on both sides of the transition, and taking into account their transition-ranks. Linear interpolation is used to join this boundary value across the transition regions of the two phonemes. An example is given in Fig. 6.3, for the second formant transition between a fricative and a vowel.

The segmental synthesis technique of Klatt (1980) is possibly the most sophisticated of the synthesis by rule systems. The method uses target values of control parameters, and a smoothing algorithm to calculate transitions. However, target values for each phoneme are not all obtained from tables; some are calculated by rules from data on their place and manner of articulation, and context. Likewise, the time constants of the parameter transitions are not stored but derived from rules on contextual information. The phoneme look-up tables are also quite elaborate. For vowels, semi-vowels and liquids, three formants and their bandwidths are tabulated for serial formant synthesis. For nasals, a pole–zero pair is added, and for stops, fricatives and affricates, six extra parameters are specified to control a parallel formant arrangement.

All these methods work rather well, and though the speech is machine-like, it can be quite intelligible. Better performance can always be obtained by storing more phoneme data and increasing the number of rules. For example, it is possible to store extra 'allophones' of some of the more variable phonemes, to be used in different contexts. This of course complicates the problem of preparing the original phonemic specification, but usually the allophonic variations can be predicted quite well, so that rules can be included to insert them automatically (Ainsworth & Millar, 1976).

Perhaps the most ambitious attempt to extend the phoneme set is that of Gagnon (1978). This scheme uses 64 'phonetic' symbols, 61 of which correspond to speech sounds. The main variation from standard IPA phonetic transcription is in the number of vowels, which includes long and short versions, and some half-diphthongs. The rules for computing the formant transitions in this scheme have not been published, but they appear to be relatively simple, since they have been incorporated into an integrated circuit, along with the synthesiser itself (Bassak, 1980).

6.5 Articulatory synthesis

In the synthesis of formant trajectories corresponding to phoneme sequences, it has been shown above that the complex relationship between articulator movements and formant transitions can be embodied in a system of rules. This is possible because, in a given language, the number of phonemes and hence the number of phoneme transitions is finite; thus the number of rules required to express them is also finite, and at most equal to the number of possible phoneme sequences. However, this dependency of formant transitions on articulator movements can be expressed more succinctly and more accurately by direct computation. This is the basis of the articulatory modelling technique proposed by Coker (1968, 1976).

This method is based on an acoustic tube model of the vocal tract, in which tube shape is represented as a set of areas at discrete points along the tract. Coker (1976) has developed a transformation from which vocal tract areas can be derived directly from positions of the articulators. This allows physiologically realistic area functions to be generated from specified articulatory gestures. These functions are then used as input to a finite difference equation (described in chapter 3), from which the resonances of the vocal tract are calculated. These resonances are formants, and if they are fed to a formant synthesiser they produce sounds corresponding to the input articulatory gestures. The formant transitions are thus 'natural', in the sense that they embody the physiological constraints of vocal tract movements.

Fig. 6.4 shows the articulatory variables proposed by Coker *et al.* (1973). They are: *X* and *Y*, tongue body height, and forward/backward movement; *B*

Fig. 6.4. The articulatory variables of the model proposed by Coker, Umeda & Browman, 1973.

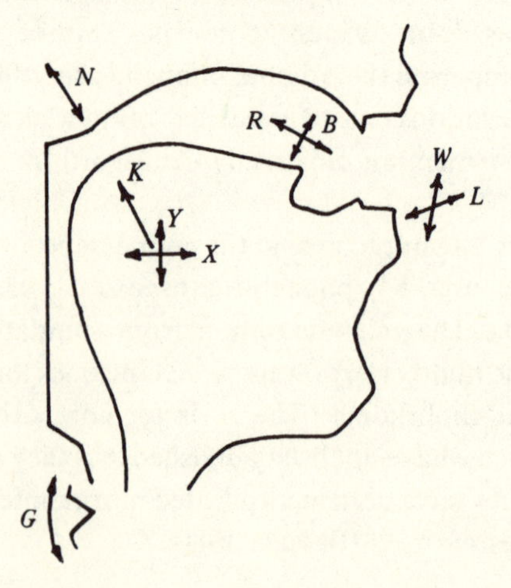

and *R*, tongue tip raising and retroflexion; *W* and *L*, lip closing and protrusion; *N*, nasal coupling; *G*, glottis position; *K*, jaw angle. These variables were deduced from a study of the dynamics of real articulatory movements, and are 'orthogonal' in the sense that their dynamics are minimally interacting. Not surprisingly, they relate rather closely to articulatory movements corresponding to individual muscle groups of the vocal system. Thus each articulatory variable may be assigned a fixed time constant. For example, *L* has time constant of about 300 ms, *W* of 50 ms, *X* and *Y* of 100 ms, and *K* of 150 ms.

Parameters to control this model take the form of step functions from one articulatory target to the next, and co-articulation comes about automatically without the use of complex rules. Rules are necessary, however, to convert target phonemes to sets of articulatory commands. The rules in this case are analogous to muscle coordination-sequences, information which is not inherent in vocal tract physiology, and which every human speaker must learn for his own particular language. This then is the difference between rules for articulatory synthesis (system model), and rules for formant synthesis (signal model); in the former co-articulation occurs as a consequence of the system dynamics, and the rules merely embody the articulatory conventions of the particular language; in the latter, the rules must encode the observed co-articulation effects at formant level, as well as the articulatory conventions.

The theoretical superiority of the articulatory model is that it can only make sounds which correspond to physiologically feasible movements of the vocal tract. Consequently co-articulation is realistic and the speech sounds more natural. Regardless of this, in practice, the rule-based formant synthesis has been preferred for economic reasons.

6.6 Concatenation of multi-phonemic units

In the synthesis of speech from phonemic specification, it is necessary to use either an elaborate computational model or a complex set of rules in order to create the effects of co-articulation without which synthetic speech sounds not only unnatural, but often unintelligible. One attempt to circumvent this problem has been to use larger, multi-phonemic units, from which to construct sentence length utterances. Both words and syllable-type units have been proposed as suitable 'building blocks' for this kind of synthesis (Rabiner, Schafer & Coker, 1971; Olive & Nakatani, 1974; Olive & Spickenagel, 1976; Fallside & Young, 1978; Levine & Sanders, 1979; Lovins, Macchi & Fujimura, 1979; Macchi, 1980; Olive, 1980; Browman, 1980).

The advantage of using such multi-phonemic units is that co-articulation effects are already inherent in their structure, and they can be used almost directly, with only minimal processing. Words are especially attractive for this application, since they are very easily strung together to make sentences, and do not require expert phonetic knowledge. However, the problem with words is that, for unrestricted synthesis, an enormous vocabulary is required. Syllable-type units, on the other hand, have the advantage that a limited (though large) set suffices to give unlimited vocabulary.

From the discussion on the nature of speech in chapter 2, it is clear that simply concatenating words, or syllable-type units, will not result in natural-sounding or even intelligible speech. There are two main problems. The first is that each unit in the utterance must be blended with its neighbours in the sequence. The second is that the whole phrase or sentence must have an appropriate prosodic stress pattern. This is more difficult to achieve, since it entails modifications to the stress of the individual units in the utterance.

The task of creating prosodic stress patterns is common to all methods of synthesising sentence length utterances, and it is such an important subject that it is convenient to postpone a detailed discussion until a later section. Assuming that an appropriate stress pattern has been generated, the problem here is how to impress it upon an utterance, built up from words or syllables, and how to blend such units together in a naturalistic manner.

A vocabulary of words or syllables can be constructed by analysis of suitable real utterances spoken in isolation, with neutral intonation. Both formants (Rabiner *et al.*, 1971) and LPC coefficients (Fallside & Young, 1978) have been used as the basic storage parameters. The pitch information (fundamental frequency) from the analysis is usually discarded, and only the resonance parameters (formants or prediction coefficients) are retained, together with amplitude and excitation data. In this form, words and syllables can be manipulated to have intonation and timing different from that in which they were originally uttered. It is the facility to do this which permits an artificial stress pattern to be superimposed on a string of concatenated units.

The concatenation problem itself is essentially one of co-articulation. Unless the stress pattern specifies a pause, the final and initial phonemes of adjacent units must be blended into each other. For word concatenation synthesis, it is fortunate that co-articulation at word boundaries is weaker than within words, so that relatively simple procedures suffice to join words together. Unfortunately, this is not so with syllables, and in fact half syllables, or 'demisyllables', are usually prefered (Lovins *et al.*, 1979).

In the word concatenation synthesis scheme of Rabiner, Schafer & Coker (1971), parameters are prepared from words spoken in isolation.

Spectral analysis is used to encode each word as resonance data. Three formants are extracted for voiced speech, and a pole–zero pair for unvoiced speech. Separate parameters are used for voiced and unvoiced amplitude, which obviates the need for a voiced/unvoiced decision. Pitch information is not stored, since this will be imposed by the prosodic stress generator. Co-articulation between words is introduced by blending the resonance control parameters at the end and beginning of adjacent units. This is done by using a set of interpolation curves, four of which are illustrated in Fig. 6.5. The appropriate interpolation curve is automatically selected by an algorithm which takes into account the values of the control parameters at the junction, and also their rates of change. The control parameter trajectories are thus smooth and continuous.

The prosodic stress generator in this system acts in two ways to modify the words in the utterance. It imposes a suitable pitch contour, and it adjusts the duration of each word, in accordance with its importance in the sentence. Unfortunately, it is not possible to expand or contract words in a linear fashion, since in real speech only vowels can be held, or abbreviated to any degree, without distorting their character. Thus the duration modification algorithm acts only on that part of the word where the spectral derivative is a minimum, that is, where the rate of change of the control parameters is least. These techniques have been used to synthesise utterances consisting of strings of digits. The result is intelligible and natural-sounding speech. The method has been applied successfully to the automatic generation of wiring instructions for the assembly of telephone equipment (Flanagan, Rabiner, Schafer & Denman, 1977).

Another word concatenation system, due to Fallside & Young (1978), uses LPC parameters to encode the words. Again, the pitch information is

Fig. 6.5. Control parameter interpolation contours, for word concatenation synthesis. From Rabiner, Schafer & Coker, 1971.

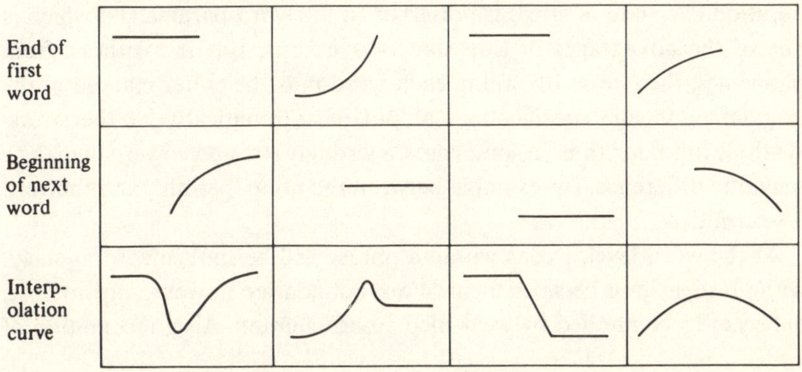

End of first word

Beginning of next word

Interp-olation curve

discarded in favour of an intonation contour imposed by the prosodic stress generator. Co-articulation between words is effected by blending the prediction coefficients at word boundaries, taking into account the type of connection required. For example, different jointing routines are used for vowel–vowel, vowel–stop, and stop–vowel junctions. This synthesis technique has been used to implement voice telemetry for a computer-controlled water supply system.

The demisyllable concatenation synthesis technique is the work of Lovins, Macchi & Fujimura (1979), and Browman (1980). The ten thousand or so syllables of the English language are split into 835 initial and final demisyllables. As well as these units, five affixes and one hundred reduced vowel demisyllables are used. All the units have been selected to be particularly easy to concatenate, and only two synthesis rules are required.

The availability of digital computers with large fast-access memory has made the concatenation of multi-phonemic units an attractive proposition. The rules in such systems are less complex than in phoneme synthesis; consequently synthesis becomes possible in real time. The technique can be viewed as being an optimal division of the segmental synthesis problem into rules and data storage. The more predictable co-articulation effects are dealt with by rules, and the more complex ones by simply storing all possible occurrences.

6.7 Synthetic stress patterns

The methods which have been devised to impress artificial stress patterns on synthetic speech are many and varied. But before attempting to describe and discuss some of them, it is worthwhile pausing to review some of the effects that stressing rules will be expected to emulate. In English, stress is essentially centred on vowels. At segmental level, one particular vowel in each word is stressed to give it prominence, and this gives the word its characteristic 'shape'. In word concatenation synthesis, the vowel duration correlate of stress is preserved in the word parameters, which is one of the advantages of this kind of synthesis. But in synthesis from phonemes, the stressed vowel in each word must be either marked in the original phonemic specification, or put in automatically by the stress-synthesising algorithm. In some cases, a wrongly stressed vowel can make a semantic difference, for example between the noun 'permit', and the verb 'to permit'.

At the word level, words within a phrase will be spoken continuously, without silent gaps between them. Word boundaries, however, are not lost, but seem to be implied by weakened co-articulation. Also, recognition of individual word 'shapes' helps to resolve ambiguities; for example, a native speaker of English would have no difficulty in distinguishing between the iso-phonemic utterances, 'white rye' and 'why try'.

At phrase level, an important question is how many words should be included in the phrase, that is, where to put the pauses. Again, there appear to be no explicit rules, except that the speaker must have sufficient breath, and that the phrase must be a semantic unit. Stress at this level is used to emphasise particular words – usually nouns and verbs. Negative stress is also possible, in that function words may be abbreviated, or reduced. Incorrect word stress within a phrase can make semantic differences; for example, the phrase 'you need another drink' can have four different meanings, depending on which word the main stress is placed on.

At sentence level, individual phrases may be emphasised within the framework of the whole utterance. Significant pauses and changes in speaking rate are two mechanisms used for this purpose. At higher levels than this, stress is still used for semantic effects. For example in the telling of stories, it has been found that words, particularly names, used for the first time are given extra stress, which gradually diminishes at each repetition thereafter (Coker, Umeda & Browman, 1973).

On a subjective level, we use the expressions 'stress', 'emphasis', 'prominence', 'intonation', etc. without implying scientific definitions of these terms. But for the purposes of the following discussion and descriptions the word 'stress' will be used in a technical sense, to denote a subjective attribute of speech, perceived to range in value from high to low. In particular, stress will be deemed to have three physical correlates: (1) fundamental frequency of voicing – pitch; (2) phoneme duration – mainly vowel elongation and reduction; (3) speech amplitude – including pauses. The task of imposing a particular stress pattern on a string of phonemes will consist in making assignments of these three variables to each relevant phoneme in the sequence.

The advent of synthetic speech has made it possible to carry out experiments, in which the variables of speech can be manipulated in a controlled and scientific manner (Umeda, 1975; Klatt, 1975; Olive, 1975). In particular, the interrelations of pitch, amplitude and duration have been investigated by Olive & Nakatani (1974). It has been shown that, within sentences, changes in pitch are the main contributors to the subjective impression of stress. Changes in phoneme duration also contribute in an important way, but fluctuations in amplitude seem to have little or no effect on perceived stress. In fact, at this level, variations in amplitude are essentially dependent on the individual phonemes in the sequence and their co-articulation. It is only at phrase level that longer term increases in amplitude are important in giving emphasis. Also, zero amplitude, that is pauses, are used to delimit larger semantic units.

Imposing a stress pattern on a single phrase can therefore be accomplished by assigning pitch and duration to each phoneme in the sequence. For this purpose the pitch contour is assumed to be continuous; however, for unvoiced segments the voicing is, of course, switched off. At the segmental level, both pitch and duration act in a contrastive manner, that is, changes in pitch from average and changes in duration from normal are the significant parameters, rather than absolute values. At higher levels, pauses, increases in amplitude and changes in average pitch may be used to delimit and emphasise phrases and sentences.

In the original speech-synthesis-by-rule system of Holmes, Mattingly & Shearme (1964), stress markers are included in the phonemic specification, to indicate the stressed syllables of the utterance. The synthesis rules then increase the duration of these phonemes by an amount specified in the look-up tables. In this first system, pitch, taken from a natural utterance, was used to supply the pitch contour. Later, Mattingly (1966), added an automatic method of assigning pitch. This uses three simple 'tone groups', one of which is allocated to each phrase on the basis of stress markers included with the phonemic specification.

In a separate development of the Holmes synthesis system, Ainsworth (1972, 1974) uses markers to indicate stressed vowels, and terminates each phrase with a full stop or a question mark. Pitch changes are automatically assigned by causing pitch to rise linearly during a stressed vowel, falling back to average about half way to the next stressed vowel. At the end of the utterance pitch is allowed to fall away for a full-stop terminator, or rise slightly for a question mark. The durations of stressed vowels are increased by an amount designed to make the time intervals between them approximately equal, in accord with the 'isochronous foot' theory of speech rhythm (Abercrombie, 1965).

The 'foot' in speech is somewhat analogous to the 'bar' in music, and syllables are analogous to notes. The isochronous foot theory requires that there be one 'beat' note, or stressed syllable, per foot, and that each foot be of approximately constant duration. Thus, feet are grouped around stressed syllables, and the duration of each syllable in a foot must be adjusted to make the total duration of each foot approximately constant.

Whitten has also used the theory of isochrony (Whitten, 1977; Whitten & Abbess, 1979) to synthesise speech from phonemic input which has two levels of stress (as well as no-stress) and normal punctuation to delimit phrases. Main (tonic) stress is reserved for the main word of each phrase, ordinary stress is allocated to the stressed syllable in each word, and no-stress is given to all other syllables. Phoneme durations are calculated by invoking the isochronous foot principle, but first an ingenious algorithm is used to divide the phoneme string into syllables, and these syllables into feet. The method of lengthening and shortening phoneme durations is quite sophisticated, and depends on internal (segmental) factors, and external (contextual) considerations. Pitch assignment is also based on the foot theory. One of five basic pitch contours (tone groups) is selected for each phrase. This tone group specifies the pitch at the first, last and tonic feet. Local perturbations are introduced, corresponding to a fall/rise in pitch for each foot in the utterance. The choice of the tone group is not automatic, but is included in the phonemic specification.

In the system of Rabiner (1968*b*, 1969), the pitch contour is based on a physiological model. This assumes that pitch is proportional to the sum of sub-glottal pressure P_s, and laryngeal tension T_L. For a phrase uttered in a single breath, both P_s and T_L will have an 'archetypal' contour, of the form shown in Fig. 6.6. For a declarative phrase, T_L is constant, but for a yes/no question, T_L rises sharply over the last 175 ms of the utterance. This gives the rising terminal pitch characteristic, required for questions.

Superimposed on this archetypal contour of P_s are perturbations due to the presence of stressed vowels and their following consonants. Approximately, a stressed vowel will raise the sub-glottal pressure by about 3 cm of water for a period of about 500 ms. The following consonant may raise or lower the pressure, depending on whether or not it is voiceless or voiced.

Fig. 6.6. Archetypal contours for sub-glottal pressure, P_s, and laryngeal tension, T_L, for a single breath group. For questions, T_L rises for the last 175 ms. After Rabiner, 1968*b*, 1969.

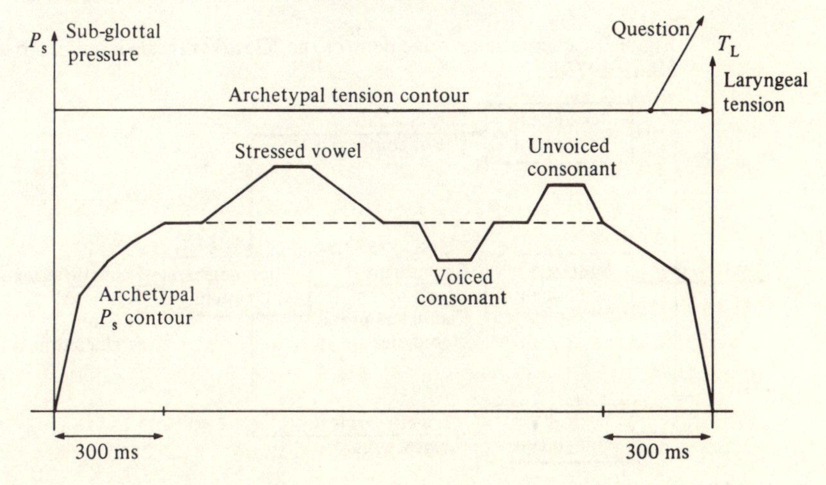

Two levels of stress are used (as well as no-stress), and these are included in the phonemic specification. Only one word per breath group receives primary stress, and only one vowel per word may receive secondary stress. All other vowels receive no-stress. Vowel duration is specified in the phoneme look-up tables, for stressed and unstressed situations.

In a later development (Rabiner, Levitt & Rosenberg, 1969), four levels of stress are used, including zero stress, together with more sophisticated rules for assigning vowel duration. Each vowel is modified by stress, in a manner depending on the stress-level, by three perturbing factors, ΔF, ΔD, and ΔC, where ΔF is a change in pitch proportional to stress-level; ΔD is an increase in vowel duration proportional to stress level, but independent of the vowel and its context; and ΔC is an increase in vowel duration independent of stress-level, but dependent on the vowel and its context. The stress levels are input to the synthesis algorithm together with the phonemic specification.

The stress assignment system of Klatt (1976*a*) views the implementation of stress as being the function of a 'phonological' program which takes into account lexical, syntactic and semantic information. This arrangement is illustrated schematically in Fig. 6.7. The various inputs to the phonological component provide the elements of the phoneme string, their lexical stress, the phrase boundaries, the speaking rate and emphasis stress. The phonological algorithm converts this information into sets of phonetic attributes, which are in turn translated into control parameters for the Klatt synthesiser described in chapter 4. Three levels of stress are used, including no-stress, which modify the basic pitch contour and also phoneme durations. The rules for calculating phoneme duration are very sophisticated, and are described at length in Klatt (1976*b*).

Fig. 6.7. Schematic arrangement of the Klatt synthesis system. After Klatt, 1976*a*.

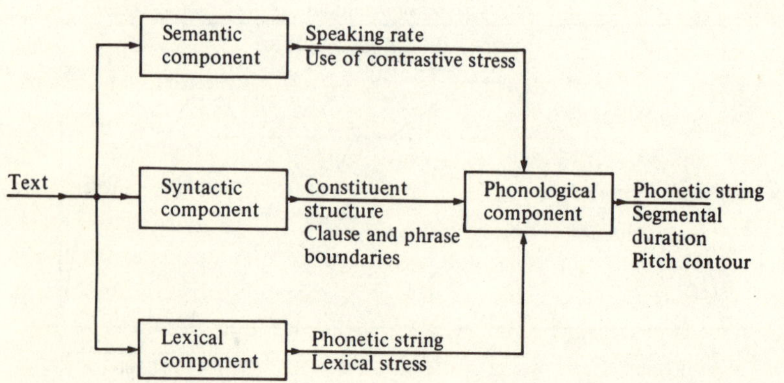

The stress assignment algorithms described so far use stress markers and phrase delimiters included in the original phonemic specification. It is obviously advantageous to be able to calculate stress patterns automatically, and attempts to do this, typically, have used rules and look-up tables. At this level, however, it is necessary to have grammatical and lexical information about every word in the proposed vocabulary. That is, a look-up table (dictionary) is required, in which each vocabulary word is listed, together with its normal stress pattern, and its part of speech. It is also necessary to embody, in the system of rules, some grammatical knowledge of the language. Since all vocabulary words must be listed, it is convenient to use the dictionary to provide a conversion from English to phonemic spelling for each word. This not only makes the construction of the dictionary much easier (phonemic listing of words is difficult), but it also enables the specification of any proposed utterance to be written in conventional orthographic text, that is, ordinary English. These 'text-to-speech' systems are described in the next section.

6.8 Text-to-speech synthesis

Text-to-speech synthesis differs in two important ways from synthesising speech via phonemic input. The first difference is simply that, since the input is now ordinary English text, with conventional punctuation, the system is much easier to use, and therefore, more widely applicable. The second difference is that since the system must calculate its own stress assignments and phrase delimiters, it must contain syntactic and linguistic knowledge, which makes it an order of magnitude more complex than phoneme-to-speech synthesis.

The first text-to-speech system was described by Teranishi & Umeda (1968), and subsequently by Umeda & Teranishi (1975). This system is based on a dictionary which lists 1500 common words. Associated with each word are its phonemic representation, its syllabification, its pronunciation and its part of speech. Each word is also catagorised as either a 'function' word or a 'content' word. Function words include conjunctions, prepositions, articles, auxiliary verbs, personal and indicative pronouns, etc. Content words are nouns, verbs, adjectives, adverbs, interrogatives, indefinite and demonstrative pronouns, etc. The program checks each input sentence for length; if it is greater than about ten syllables, it subdivides it using syntactic information, until each division or breath group is within the required limit. Two stress levels are used (stress and no-stress), the rule being that function words are not stressed and content words have one stressed syllable. Pitch and vowel duration are modified according to rules similar to those already described in the previous section.

In a system developed by Ainsworth (1973), the English spelling problem is solved by using a combination of dictionary and rules. Words input to the system are first checked against a list of exceptions; if they are not on the list, then spelling rules are applied, which take account of each letter and its context. Words that are on the exception list have their phoneme equivalent and stress pattern given directly. The spelling rules also include an algorithm for allocating stress to the appropriate vowel. The rule is that stress is always assigned to the first syllable, except when that syllable is a prefix, in which case stress is displaced to the second syllable. Again all words are classified as either function or content words, but in this system all function words are included in the exceptions list. Thus any word not on the list is assumed to be a content word. Grammatical categories are given to words on the list, so that a simple syntax analysis is possible. This is used to subdivide the input text into breath groups. The boundaries between breath groups are located by a simple algorithm. Breath group boundaries are placed at the following, in order of priority: (*a*) a punctuation mark; (*b*) before a conjunction; (*c*) between a noun phrase and a verb phrase; (*d*) before a prepositional phrase; (*e*) before a noun phrase; and (*f*) after a fixed number of phonemes have been included in the breath group. Boundaries are inserted sequentially by this process, until each breath group is within the maximum of 50 phonemes. The stress and phrase limits derived by this procedure are converted to pitch and vowel duration assignments, using the methods described previously, and in Ainsworth (1972, 1974).

Perhaps the most comprehensive text-to-speech system has been described by Coker, Umeda & Browman (1973), which has been developed to drive the Coker articulatory synthesiser. Text is divided into breath groups using a grammatical-category transition matrix. This is a table which assigns a pause probability for transitions between words of the various grammatical categories. Thus a pause potential is inserted after every word of the input text, and subdivision into breath groups is effected using these potentials together with a maximum length criterion. Various types of pauses are distinguished, including a 'pseudo-pause', in which phonemes on either side are modified in the usual way, but no actual silent gap is inserted. Stress is assigned to every word in the breath group according to its grammatical significance. A nine-point stress scale, illustrated in Fig. 6.8, is suggested. Vowel duration is modified by use of the formula

$$\text{duration in ms} = 80 + 130(1 + C)S$$

where S is a 'stress situation factor', and C is a factor which takes account of the following consonant. The stress situation factor is designed to include the effects of word type and context. Pitch is treated as an articulatory variable, so that target values of pitch are assigned to each phoneme along the lines of the stress scheme shown in Fig. 6.8. This sequence of target pitch values is smoothed by a characteristic filter in the same way as the other articulatory control signals. Some allophonic variations are automatically taken care of by the articulatory dynamics. Other allophones are inserted by the rules, which take account of context. The speech output from this system is quite natural-sounding, and plausible in its prosodic structure.

The word concatenation synthesis system of Fallside & Young (1978) is unusual in that the utterance to be synthesised is itself generated by computer program. This is called 'synthesis from concept' (Young & Fallside, 1979), and is designed to permit information from a data-base to be presented in verbal form. The utterance generated by the 'concept-to-text' algorithm is prepared for synthesis as speech by an analysis of its syntactic structure (Young & Fallside, 1980). Since the grammatical form of the utterance is synthesised by the concept-to-text program, there are no grammatical ambiguities to be dealt with. The analysis of a typical sentence

Fig. 6.8. Nine-level prosodic stress scheme. From Coker, Umeda & Browman, 1973.

Stress level	Example	Boundary consonant elongation in ms	Main vowel pitch in Hz
1	New nouns	60	160
2	Prepositions used as complements	60	160
3	New infrequent verbs; adjectives, adverbs; repeated nouns	60	150
4	Repeated infrequent verbs	60	140
5	Interrogatives, quantitatives	50	132
6	Frequent verbs	40	125
7	Less frequent function words	25	120
8	Ordinary function words	10	115
9	'schwa' function words	0	100

is illustrated in Fig. 6.9. The utterance is divided into word groups by use of these syntactic divisions and a maximum length criterion. In the example of Fig. 6.9, the first word group boundary is placed at point 1, because this is the primary syntactic division of the sentence. Since both subdivisions still exceed the maximum length of 24 syllables, they are subdivided again. This is accomplished by breaks at points 2 and 3. Again, point 2 is placed at a major syntactic division. Point 3, however, is not placed after the word 'is', because of a rule which forbids function words being split off as word groups. Instead, point 3 is displaced to the next most important syntactic break.

Stress is assigned automatically, using a three-level scale. Word stress is obtained from the vocabulary, and extra, phrase stress is given to the last content word in the word group. The stress pattern is then adjusted, in an attempt to alternate stressed and unstressed syllables. This stress assignment is translated directly into increases in syllable duration, and the time intervals between stressed syllables are made approximately equal by further modifications of syllable length. Essentially, short feet are lengthened by elongating the syllables of content words, and long feet are shortened by reducing the syllables of function words. A pitch contour is imposed on each word group, by automatic selection of one of three standard tone groups. Local variations in pitch are then impressed upon this basic contour in synchronism with the syllables of the utterance. The actual shape of these syllable pitch-patterns is dependent on the level of stress and the context of the syllable.

Fig. 6.9. Example of the syntactic analysis of a sentence, and its subdivision into word groups. After Young & Fallside, 1980.

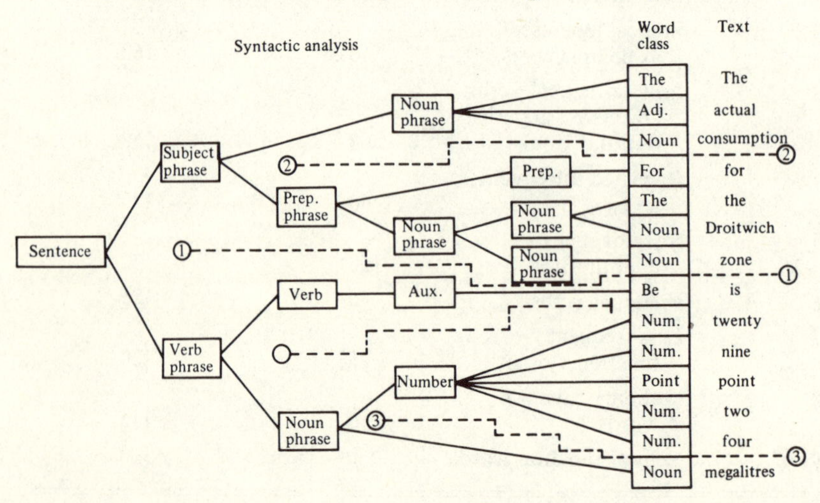

The synthesis from text system of Allen (1973, 1976) is probably the most ambitious of its kind. Its objective is to synthesise speech from unrestricted English text, that is, it must be able to deal with every word in the English language. As Allen observes, language is an organic, changing system of representation, and simply storing all possible words in a vast dictionary is neither desirable nor practical. Instead, Allen proposes a combined system of rules, and a dictionary of word roots or 'morphs'. Every word is regarded as either a morph (for example, search, light, etc.), or a compound morph (searchlight), or a morph plus affix (research, searched, enlighten, lighting, lightening, etc.). The basis of the rules system is that prefixes and suffixes can be stripped away from any complex word, to reveal its essential morph structure. An affix-stripping algorithm has been developed to do this (Lee, 1967). An input word is first presented to the morph dictionary; if it is not listed, then it is put into the affix-stripping algorithm, and, if possible, an affix is removed. The reduced word is then checked against the morph list; if it is not found then another affix is stripped away. This sequence is repeated until the reduced word is identified as a morph. Or, if the affix-stripping algorithm fails, and the reduced word is not listed, letter-to-sound rules are applied to 'guess' its pronunciation.

The affix-stripping algorithm considerably reduces the number of words to be stored in the dictionary, and in addition provides semantic and syntactic information about the function of the words. For example, the suffix 'ness' forms nouns from adjectives, as in the word 'kindness'. This not only gives the correct grammatical category, noun, but also denotes which of the two meanings of 'kind' is intended. Because of the potential ambiguities inherent in multi-affixed words, compatibility rules are included in the algorithm. Rules are also used to take care of those cases in which the form of the morph changes when taking an affix. For example, 'intension' is 'intend + ion' not 'intens + ion'. In deducing the rules and constructing the morph dictionary, Allen processed over a million words of text. The resulting dictionary contains about 10 000 morphs, which in conjunction with the rule system is capable of dealing with more than 100 000 different words. This text analysis scheme has been used to provide the semantic, syntactic and lexical input to the Klatt phonological and segmental synthesis algorithms. The result is highly intelligible synthetic speech, though still with an unmistakable machine accent.

References

Abercrombie, D. (1965). *Studies in Phonetics and Language.* London: Oxford University Press.

Ainsworth, W.A. (1972). A real-time speech synthesiser. *Institute of Electrical and Electronic Engineers, Trans.* **AU-20**, 397–9.

Ainsworth, W.A. (1973). A system for converting English text into speech, *Institute of Electrical and Electronic Engineers, Trans.* **AU-21**, 288–90.

Ainsworth, W.A. (1974). Performance of a speech synthesis system. *International Journal of Man–Machine Studies*, **6**, 493–511.

Ainsworth, W.A. & Millar, J.B. (1976). Allophonic variation of stop consonants in a speech synthesis-by-rule program. *International Journal of Man–Machine Studies*, **8**, 159–68.

Allen, J. (1973). Reading machines for the blind: The technical problems and the methods adopted for their solution. *Institute of Electrical and Electronic Engineers, Trans.* **AU-21**, 259–64.

Allen, J. (1976). Synthesis of speech from unrestricted text. *Institute of Electrical and Electronic Engineers, Proc.*, **64**, 433–42.

Atal, B.S. (1970). Speech analysis and synthesis by linear prediction of the speech wave. *Journal of the Acoustical Society of America*, **47**, 65(A).

Atal, B.S. & David, N. (1979). On synthesising natural sounding speech by linear prediction. *Institute of Electrical and Electronic Engineers*, **ICASSP-79**, 44–7.

Atal, B.S. & Hanauer, S.L. (1971). Speech analysis and synthesis by linear prediction of the speech wave. *Journal of the Acoustical Society of America*, **50**, 637–55.

Bassak, G. (1980). Phoneme based speech chip needs less memory. *Electronics* (USA), **53**, 43–4.

Bell, C.G., Fujisaki, H., Heinz, J.M., Stevens, K.N. & House, A.S. (1961). Reduction of speech spectra by analysis-by-synthesis techniques. *Journal of the Acoustical Society of America*, **33**, 1725–36.

Browman, C.P. (1980). Rules for demisyllable synthesis using LINGUA, a language interpreter. *Institute of Electrical and Electronic Engineers*, **ICASSP-80**, 561–64.

Burg, J.P. (1975). Maximum entropy spectral analysis. *Proceedings of the 37th Meeting of the Society of Exploration Geophysicists.* Houston, Texas: *Geophysics.*

Cervantes (Saavedra), M. de (1970). *Don Quixote*, Part Two Tr. J.M. Cohen, Chapter Sixty-Two 'The Enchanted Head'. Harmondsworth: Penguin.

Chandra, S. & Lin, W.C. (1974). Experimental comparisons between stationary and non-stationary formulations of linear prediction. *Institute of Electrical and Electronic Engineers, Trans.* **ASSP-22**, 403–15.

Chandra, S. & Lin, W.C. (1977). Linear prediction with a variable analysis frame size. *Institute of Electrical and Electronic Engineers, Trans.* **ASSP-25**, 322–30.

Chong Kwan Un (1978). A low-rate digital formant vocoder. *Institute of Electrical and Electronic Engineers, Trans.* **COM-26**, 344 –55.

Chong Kwan Un & Magill, D.T. (1975). The residual-excited linear prediction vocoder with transmission rate below 9.6 kb/s. *Institute of Electrical and Electronic Engineers, Trans.* **COM-23**, 1466–74.

Coker, C.H. (1963). Computer simulated analyser for a formant vocoder. *Journal of the Acoustical Society of America*, **35**, 1911(A).

Coker, C.H. (1965). Real-time formant vocoder, using a filter bank, a general purpose computer and an analog synthesiser. *Journal of the Acoustical Society of America*, **38**, 940(A).

Coker, C.H. (1967). Synthesis by rule from articulatory parameters. *Institute of Electrical and Electronic Engineers, Conference on Speech Communication Processes*, Paper A9, pp. 52–3.

Coker, C.H. (1968). Speech synthesis with a parametric articulatory model. In Flanagan & Rabiner (1973), pp. 135–9.

Coker, C.H. (1976). A model of articulatory dynamics and control. *Institute of Electrical and Electronic Engineers, Proc.*, **64**, 452–60.

Coker, C.H., Umeda, N. & Browman, C.P. (1973). Automatic synthesis from ordinary English text. *Institute of Electrical and Electronic Engineers, Trans.* **AU-21**, 293–8.

Cooper, F.S., Delattre, P.C., Liberman, A.M., Borst, J.M. & Gerstman, L.J. (1959). Some experiments on the perception of synthetic speech sounds. *Journal of the Acoustical Society of America*, **24**, 597–606.

Cooper, F.S., Liberman, A.M. & Borst, J.M. (1951). The interconversion of audible and visible patterns as a basis for research in speech perception. *Proc. National Academy of Science*, **37**, 318–25.

Cowie, R.I.D. & Douglas-Cowie, E. (1983). Speech production in profound postlingual deafness. In *Hearing science and hearing disorders*, Ed. M.E. Lutman & M.P. Haggard. London: Academic Press.

Dixon, N.R. & Maxey, H.D. (1968). Terminal analog synthesis of continuous speech using the diphone method of segment assembly. *Institute of Electrical and Electronic Engineers, Trans.* **AU-16**, 40–50.

Dudley, H. (1936). Synthesising speech. *Bell Labs Record*, **15**, 98–102.

Dudley, H. (1939*a*). Remaking speech. *Journal of the Acoustical Society of America*, **11**, 169–77.

Dudley, H. (1939*b*). The vocoder. *Bell Labs Record*, **17**, 122–6.

Dudley, H. (1940), The carrier nature of speech. *Bell System Technical Journal*, **19**, 459–515.

Dudley, H. (1955). Fundamentals of speech synthesis. *Journal of Audio Engineers Society*, **3**, 170–85.

Dudley, H., Riesz, R.R. & Watkins, S.A. (1939). A synthetic speaker. *Journal of the Franklin Institute*, **227**, 739–64.

Dunn, H.K. (1950). The calculation of vowel resonances, and an electrical vocal tract. *Journal of the Acoustical Society of America*, **22**, 740–53.

Dunn, H.K., Flanagan, J.L. & Gestrin, P.J. (1962). Complex zeros of a triangular approximation to the glottal wave. *Journal of the Acoustical Society of America*, **34**, 1977(A).

Estes, S.E., Kerby, H.R., Maxey, H.D. & Walker, R.M. (1964). Speech synthesis from stored data. *IBM Journal of Research and Development*, **8**, 2–12.

Fallside, F. & Young, S.J. (1978). Speech output from a computer-controlled water supply network. *Institute of Electrical and Electronic Engineers, Proc.*, **125(2)**, 157–61.

Fant, G. (1956). On the predictability of formant levels and spectrum envelopes from formant frequencies. In *For Roman Jacobson*, pp. 109–20. 's-Gravenhage: Mouton.

Fant, G. (1960). Acoustic theory of speech production. 's-Gravenhage: Mouton.

Fant, G. (1973). *Speech Sounds and Features*. Cambridge, Massachusetts: MIT Press.

Flanagan, J.L. (1956). Automatic extraction of formant frequencies from continuous speech. *Journal of the Acoustical Society of America*, **28**, 110–18.

Flanagan, J.L. (1957). Note on the design of 'Terminal Analog' speech synthesisers. *Journal of the Acoustical Society of America*, **29**, 306–10.

Flanagan, J.L., (1960). Resonance-vocoder and baseband complement. *Institute of Electrical and Electronic Engineers, Trans.* **AU-8**, 95–102.

Flanagan, J.L. (1972). *Speech synthesis, analysis and perception* New York: Springer-Verlag.

Flanagan, J.L., Coker, C.H. & Bird C.M. (1962). Computer simulation of a formant vocoder synthesiser. *Journal of the Acoustical Society of America*, **35**, 2003(A).

Flanagan, J.L., Coker, C.H., Rabiner, L.R., Schafer, R.W. & Umeda, N. (1970). Synthetic voices for computers. *Institute of Electrical and Electronic Engineers, Spectrum*, **7**, 22–45.

Flanagan, J.L. & House, A.S. (1956). Development and testing of a formant-coding speech compression system. *Journal of the Acoustical Society of America*, **28**, 1099–1106.

Flanagan, J.L. & Ishizaka, K. (1976). Automatic generation of voiceless excitation in a vocal cord-tract speech synthesiser. *Institute of Electrical and Electronic Engineers Trans.* **ASSP-24**, 163–70.

Flanagan, J.L. & Ishizaka, K. (1978). Computer model to characterise the air volume displaced by the vibrating vocal cords. *Journal of the Acoustical Society of America*, **63**, 1559–65.

Flanagan, J.L., Ishizaka, K. & Shipley, K.L. (1975). Synthesis of speech from a dynamic model of the vocal cords and vocal tract. *Bell System Technical Journal*, **54**, 485–505.

Flanagan, J.L. & Landgraf, L.L. (1968). Self-oscillating source for vocal-tract synthesis. *Institute of Electrical and Electronic Engineers Trans.* **AU-16**, 57–64.

Flanagan, J.L. & Rabiner, L.R. (Eds.) (1973). *Speech Synthesis.* Stroudsburg, Pennsylvania: Dowden, Hutchinson & Ross.

Flanagan, J.L., Rabiner, L.R., Schafer, R.W. & Denman, J.D. (1977). Wiring telephone apparatus from computer generated speech. *Bell System Technical Journal*, **51**, 391–7.

Frantz, G. & Kun-Shan Lin (1982). Speech technology in consumer products. *Speech Technology* (USA), **1**, 25–34.

Gagnon, R.T. (1978). Votrax real-time hardware for phoneme synthesis of speech. *Institute of Electrical and Electronic Engineers*, **ICASSP-78**, 175–8.

Gold, B. & Rabiner L.R. (1968). Analysis of digital and analog formant synthesisers. *Institute of Electrical and Electronic Engineers, Trans.* **AU-16**, 81–94.

Gold, B. & Rader, C.M. (1967). The channel vocoder. *Institute of Electrical and Electronic Engineers, Trans.* **AU-15**, 148–61.

Goldberg, A.J. & Ross, M.J. (1973). Implementation of a linear predictive vocoder. *Journal of the Acoustical Society of America*, **54**, 340(A).

Goldberg, A.J. & Shaffer, H.L. (1975). A real-time adaptive predictive coder using small computers. *Institute of Electrical and Electronic Engineers, Trans.* **COM-23**, 1443–51.

Golden, R.M. (1963). Digital computer simulation of a sampled-data, voice-excited vocoder. *Journal of the Acoustical Society of America*, **35**, 1358–66.

Gopinath, B. & Sondhi, M.M. (1970). Determination of the shape of the human vocal tract from acoustical measurements. *Bell System Technical Journal*, **49**, 1195–214.

Green, N. (1976). Analysis-synthesis using pole-zero approximations to speech spectra. *Institute of Electrical and Electronic Engineers*, **ICASSP-76**, 306.

Hofstetter, E.M., Tierney, J. & Wheeler, O. (1977). Microprocessor realisation of a linear predictive vocoder. *Institute of Electrical and Electronic Engineers, Trans.* **ASSP-25**, 379–87.

Holmes, J.N. (1972). *Speech synthesis.* London: Mills & Boon.

Holmes, J.N. (1973). The influence of glottal waveform on the naturalness of speech from a parallel formant synthesiser. *Institute of Electrical and Electronic Engineers, Trans.* **AU-21**, 298–305.

Holmes, J.N., Mattingly, I.G. & Shearme, J.N. (1964). Speech synthesis by rule. *Language and Speech* **7**, 127–43.

Ishizaka, K. & Flanagan, J.L. (1972). Synthesis of voiced sounds from a two-mass model of the vocal cords. *Bell System Technical Journal*, **51**, 1233–68.

Itakura, F. & Saito, S. (1968). Analysis synthesis telephony based on the maximum likelihood method. In Flanagan & Rabiner (1973), pp. 287–92.

Itakura, F. & Saito, S. (1970). A statistical method for estimation of speech spectral density and formant frequencies. *Electronics and Communication*, Japan, **53-A**, 36–43.

Itakura, F. & Saito, S. (1972). On the optimum quantization of feature parameters in the PARCOR speech synthesiser. In Flanagan & Rabiner (1973), pp. 301–4.

Itakura, F., Saito, S., Koike, T., Sawabeh, H. & Nishikawa, M. (1972). An audio response unit based on partial autocorrelation. *Institute of Electrical and Electronic Engineers, Trans.* **COM-20**, 792–7.

Kelly, J.L. Jr & Gerstman, L.J. (1961). An artificial talker driven from a phonetic input. *Journal of the Acoustical Society of America*, **33**, 835(A).

Kelly, J.L. Jr & Lochbaum, C.C. (1962). Speech synthesis. In Flanagan & Rabiner (1973). pp. 127–30.

Kelly, L.C. (1970). Speech and vocoders. *Radio Electron. Eng.* (GB), **40**, 73–82.

Keltai, M.A. (1980). Electronic simulation of the vocal tract. M.Sc. thesis, Queen's University, Belfast.

von Kempelen. (1791) *The mechanism of speech and the construction of a speaking machine*. Vienna.

Klatt, D.H. (1975). Vowel lengthening is syntactically determined in a discourse. *Journal of Phonetics*, **3**, 161–72.

Klatt, D.H. (1976*a*). Structure of a phonological rule component for a synthesis-by-rule program. *Institute of Electrical and Electronic Engineers, Trans.* **ASSP-24**, 391–8.

Klatt, D.H. (1976*b*). Linguistic use of segmental duration in English: Acoustic and perceptual evidence. *Journal of the Acoustical Society of America.* **59**, 1208–21.

Klatt, D.H. (1980). Software for a cascade/parallel formant synthesiser. *Journal of the Acoustical Society of America*, **67**, 971–95.

Ladefoged, P., Harshman, R., Goldstein, L. & Rice, L. (1978). Generating vocal tract shapes from formant frequencies. *Journal of the Acoustical Society of America*, **64**, 1027–35.

Lawrence, W. (1953). The synthesis of speech from signals which have a low information rate. In *Communication theory*, pp. 460–9. London: Butterworth.

Lee, F.F. (1967). Automatic grapheme-to-phoneme translation of English. *Journal of the Acoustical Society of America*, **41**, 1594.

Levine, A. & Sanders, W.R. (1979). The MISS speech synthesis system. *Institute of Electrical and Electronic Engineers*. **ICASSP-79**, 899–902.

Levinson, S.E., Coker, C.H. & Schmidt, C.E. (1981). Adaptive emulation of human speech by an articulatory speech synthesiser. Bell Laboratories, Jan. 1981.

Liberman, A., Ingermann, F., Lisker, L., Delattre, P. & Cooper, F. (1959). Minimal rules for synthesising speech. *Journal of the Acoustical Society of America*, **31**, 1490–99.

Kun-Shan Lin. and Ying L. Tsui. (1981). LPC compressed speech at 850 bits per second. *Institute of Electrical and Electronic Engineers*, **ICASSP-81**, 5–7.

Linggard, R. & Marlow, F.J. (1979). A programmable, digital speech synthesiser. *IEE Journal on Computers and Digital Techniques* (GB), **2**, 191–6.

Longuet-Higgins, H.C. (1978). The pole-zero analysis of speech by generalised linear prediction theory. Proceedings of British Institute of Acoustics Conference, 1978. Available from the British Institute of Acoustics.

Lovins, J.B., Macchi, M.J. & Fujimura, O. (1979). A demisyllable inventory for speech synthesis. *Journal of the Acoustical Society of America*, **65**, 5130–31.

Macchi, M.J. (1980). A phonetic dictionary for demisyllable speech synthesis. *Institute of Electrical and Electronic Engineers*, **ICASSP-80**, 565–7.

Makhoul, J. (1975*a*). Spectral linear prediction: Properties and applications. *Institute of Electrical and Electronic Engineers, Trans.* **ASSP-23**, 283–96.

Makhoul, J. (1975*b*). Linear prediction – a tutorial review. *Proceedings Institute of Electrical and Electronic Engineers*, **63**, 561–80.

Makhoul, J. (1977). Stable and efficient lattice methods for linear prediction. *Institute of Electrical and Electronic Engineers, Trans.* **ASSP-25**, 423–8.

Markel, J.D. (1972*a*). Digital inverse filtering – a new tool for formant trajectory estimation. *Institute of Electrical and Electronic Engineers, Trans.* **AU-20**, 129–37.

Markel, J.D. (1972*b*). The SIFT algorithm for fundamental frequency estimation. *Institute of Electrical and Electronic Engineers, Trans.* **AU-20**, 367–77.

Markel, J.D. (1973). Application of a digital inverse filter for automatic formant and F_0 analysis. *Institute of Electrical and Electronic Engineers, Trans.* **AU-21**, 154–60.

Markel, J.D. & Gray, A.H., Jr (1973). On autocorrelation equations as applied to speech analysis. *Institute of Electrical and Electronic Engineers, Trans.* **AU-21**, 69–79.

Markel, J.D. & Gray, A.H., Jr (1974). A linear prediction vocoder simulation based upon the autocorrelation method. *Institute of Electrical and Electronic Engineers, Trans.* **ASSP-23**, 124–34.

Markel, J.D. & Gray, A.H., Jr (1976). *Linear prediction of speech*. New York: Springer-Verlag.

Mattingly, I.G. (1966). Synthesis by rule of prosodic features. *Language and Speech*, **9(1)**, 1–13.

McCandless, S.S. (1974). An algorithm for automatic formant extraction using linear prediction spectra. *Institute of Electrical and Electronic Engineers, Trans.* **ASSP-22**, 135–41.

McGurk, H. & MacDonald, J. (1976). Hearing lips and seeing voices. *Nature* (GB), **264**, 746–8.

Mermelstein, P. (1971). Calculation of the vocal tract transfer function for speech synthesis applications. In Flanagan & Rabiner (1973), pp. 131–4.

Morf, M. *et al.* (1977). Efficient solution of covariance equations for linear prediction. *Institute of Electrical and Electronic Engineers, Trans.* **ASSP-25**, 429–33.

Morgan, D.R. (1975). A note on real-time linear prediction of speech waveforms. *Institute of Electrical and Electronic Engineers, Trans.* **ASSP-23**, 386–7.

Olive, J.P. (1975). Fundamental frequency rules for the synthesis of simple declarative English sentences. *Journal of the Acoustical Society of America*, **57**, 476–82.

Olive, J.P. (1980). A scheme for concatenating units for speech synthesis. *Institute of Electrical and Electronic Engineers*, **ICASSP-80**, 568–71.

Olive, J.P. & Nakatani, L.H. (1974). Rule-synthesis of speech by word concatenation: A first step. *Journal of the Acoustical Society of America*, **55**, 660–66.

Olive, J.P. & Spickenagel, N. (1976). Speech resynthesis from phoneme-related parameters. *Journal of the Acoustical Society of America*, **59**, 993.

Paget, R. (1930). *Human speech.* London: Routledge and Kegan Paul Ltd.

Paige, A. & Zue, V.W. (1970). Computation of vocal tract area functions. *Institute of Electrical and Electronic Engineers Trans.* **AU-17**, 7–18.

Paul, A.P., House, A.S. & Stevens, K.N. (1964). Automatic reduction of vowel spectra; an analysis by synthesis method and its evaluation. *Journal of the Acoustical Society of America*, **36**, 303–8.

Perkell, J.D.S. (1981). On the use of feedback in speech production. In *The Cognitive representation of speech.* Ed. T. Myers, J. Laver & J. Anderson, Amsterdam: North Holland.

Pinson, E.N. (1963). Pitch synchronous time-domain estimation of formant frequencies and bandwidths. *Journal of the Acoustical Society of America*, **35**, 1264–73.

Rabiner, L.R. (1968*a*). Digital-formant synthesiser for speech synthesis studies. *Journal of the Acoustical Society of America*, **43**, 822–8.

Rabiner, L.R. (1968*b*). Speech synthesis by rule: An acoustic domain approach. *Bell System Technical Journal*, **47**, 17–37.

Rabiner, L.R. (1969). A model for synthesising speech by rule. *Institute of Electrical and Electronic Engineers, Trans.* **AU-17**, 7–13.

Rabiner, L.R., Atal, B.S. & Sambur, M.R. (1977). LPC prediction error-analysis and its variation with the position of the analysis frame. *Institute of Electrical and Electronic Engineers, Trans.* **ASSP-25**, 434–42.

Rabiner, L.R. & Gold, B. (1975). *Theory and application of digital signal processing.* New York: Prentice-Hall.

Rabiner, L.R., Levitt, H. & Rosenberg, A.E. (1969). Investigation of stress patterns for speech synthesis by rule. *Journal of the Acoustical Society of America*, **45**, 92–101.

Rabiner, L.R. & Schafer, R.W. (1976). Digital techniques for computer voice response: Implementation and applications. *Proceedings, Institute of Electrical and Electronic Engineers*, **64**, 416–33.

Rabiner, L.R. & Schafer, R.W. (1980). *Digital processing of speech signals*, p. 404 Englewood Cliffs, New Jersey: Prentice-Hall.

Rabiner, L.R., Schafer, R.W. & Coker, C.H. (1971). Digital hardware for speech synthesis. *Institute of Electrical and Electronic Engineers, Trans.* **COM-19**, 1016–20.

Rosen, G. (1956). Dynamic analog speech synthesiser. *Journal of the Acoustical Society of America*, **30**, 201–9.

Rosenberg, A.E. (1971). Effect of glottal pulse shape on the quality of natural vowels. *Journal of the Acoustical Society of America*, **49**, 583–90.

Sambur, M.R. (1975). An efficient linear prediction vocoder. *Bell System Technical Journal*, **54**, 1693–1723.

Sambur, M.R., Rosenberg, A.E., Rabiner, L.R. & McGonegal, C.A. (1978). On reducing the buzz in LPC synthesis. *Journal of the Acoustical Society of America*, **63**, 918–24.

Samei, T., Asada, A. & Nakata, K. (1980). High quality PARCOR speech synthesiser. *Institute of Electrical and Electronic Engineers, Trans.* **CE-26**, 353–9.

Schafer, R.W. & Rabiner, L.R. (1970). System for automatic formant analysis of voiced speech. *Journal of the Acoustical Society of America*, **47**, 634–48.

Schroeder, M.R. (1956). On the separation and measurement of formant frequencies. *Journal of the Acoustical Society of America*, **28**, 159(A).

Schroeder, M.R. (1966). Vocoders: Analysis and synthesis of speech. *Proceedings of the Institute of Electrical and Electronic Engineers*, **54**, 720–34.

Schroeder, M.R. & David, E.E., Jr. (1960). A vocoder for transmitting 10 kc/s speech over a 3.5 kc/s channel. *Acoustica* **10**, 35–43.

Seeviour, P.M., Holmes, J.N. & Judd, M.W. (1976). Automatic generation of control signals for a parallel formant speech synthesiser. *Institute of Electrical and Electronic Engineers* **ICASSP-76**, 690–93.

Smith, C.P. (1963). Voice-communications method using pattern matching for data compression. *Journal of the Acoustical Society of America*, **35**, 805(A).

Stansfield, E.V. (1973). Determination of vocal tract area function from transfer impedance. *Proceedings British Institute of Electrical Engineers*, **120**, 153–8.

Stevens, K.N., Bastide, R.P. & Smith, C.P. (1960). Electrical synthesiser of continuous speech. *Journal of the Acoustical Society of America*, **32**, 47–55.

Stevens, K.N., Kasowski, S. & Fant, G. (1953). An electrical analog of the vocal tract. *Journal of the Acoustical Society of America*, **25**, 734–42.

Stewart, J.Q. (1922). An electrical analogue of the vocal organs. *Nature* (GB), **110**, 311–12.

Strube, H. (1974). Determination of the instant of glottal closure from the speech wave. *Journal of the Acoustical Society of America*, **56**, 1625–9.

Strube, H.W. (1977). Analogue discrete-time filter for speech synthesis. *Institute of Electrical and Electronic Engineers, Trans.* **ASSP-25**, 50–55.

Suzuki, J., Kadokawa, Y. & Nakata, K. (1963). Formant frequency extraction by the method of moment calculation. *Journal of the Acoustical Society of America*, **35**, 1345–53.

Teranishi, R. & Umeda, N. (1968). Use of a pronouncing dictionary in speech synthesis experiments. In Flanagan & Rabiner (1973), pp. 412–15.

Umeda, N. (1975). Vowel duration in American English. *Journal of the Acoustical Society of America*, **58**, 434–45.

Umeda, N. & Teranishi, R. (1975). The parsing program for automatic text-to-speech synthesis, developed at the Electrotechnical Laboratory in 1968. *Institute of Electrical and Electronic Engineers, Trans.* **ASSP-23**, 183–8.

Underwood, M.J. & Martin, M.J. (1976). A multi-channel formant synthesiser for computer speech response. Proceedings British Institute of Acoustics Autumn Conference. Available from the British Institute of Acoustics.

Viswanathan, R. & Makhoul, J. (1975). Quantization properties of transmission parameters in linear predictive systems. *Institute of Electrical and Electronic Engineers, Trans.* **ASSP-23**, 309–20.

Wakita, H. (1973). Direct estimation of the vocal tract shape by inverse filtering of the acoustic speech waveform. *Institute of Electrical and Electronic Engineers, Trans.* **AU-21**, 417–27.

Weibel, E.S. (1955). Vowel synthesis by resonant circuits. *Journal of the Acoustical Society of America*, **27**, 858–65.

Whitten, I.H. (1977). A flexible scheme for assigning timing and pitch to synthetic speech. *Language and Speech*, **20**, 240–60.

Whitten, I.H. & Abbess, J. (1979). A microcomputer based speech synthesis system. *International Journal of Man–Machine Studies*, **11**, 586–620.

Wiggins, R.H. (1978). Low cost speech synthesis. *1978 MIDCON Technical Papers*, 16/3/1–3. North Hollywood, California: Western Periodicals.

Wong, D.Y. & Markel, J.D. (1978). An excitation function for LPC synthesis which retains the human glottal pulse characteristics. *Institute of Electrical and Electronic Engineers* **ICASSP-78**, 171–4.

Young, S.J. & Fallside, F. (1979). Speech synthesis from concept: A method for speech output from information systems. *Journal of the Acoustical Society of America*, **66**, 685–95.

Young, S.J. & Fallside, F. (1980). Synthesis by rule of prosodic features in word concatenation synthesis. *International Journal of Man–Machine Studies*, **12**, 241–58.

Index